CELL BIOLOGY RESEARCH PROGRESS

CRYOPRESERVATION

TECHNOLOGIES, APPLICATIONS AND RISKS/OUTCOMES

Cell Biology Research Progress

Additional books in this series can be found on Nova's website under the Series tab.

Additional e-books in this series can be found on Nova's website under the e-book tab.

Microbiology Research Advances

Additional books in this series can be found on Nova's website under the Series tab.

Additional e-books in this series can be found on Nova's website under the e-book tab.

CELL BIOLOGY RESEARCH PROGRESS

CRYOPRESERVATION

TECHNOLOGIES, APPLICATIONS AND RISKS/OUTCOMES

ACEL COLVERT
AND
HUGH COTY
EDITORS

NOVA BIOMEDICAL

New York

For permission to use material from this book please contact us:
Telephone 631-231-7269; Fax 631-231-8175
Web Site: http://www.novapublishers.com

NOTICE TO THE READER

Additional color graphics may be available in the e-book version of this book.

LIBRARY OF CONGRESS CATALOGING-IN-PUBLICATION DATA

ISBN: 978-1-62618-474-9

Published by Nova Science Publishers, Inc. † New York

CONTENTS

Preface		**vii**
Chapter 1	Cryopreservation of Filamentous Fungi: A Review *Ladislav Homolka*	**1**
Chapter 2	Differential Scanning Calorimetry: A Useful Tool for Better Understanding Crystallisation and Slow-freezing Protocols *P. Bruyere, A. Baudot, P. Guérin, G. Louis, T. Joly and S. Buff*	**67**
Chapter 3	Alternatives to Cryopreservation for the Short and Long-Term Storage of Mammalian Cells *Karen Coopman*	**91**
Chapter 4	Mathematical Modeling and Simulations in Cryopreservation of Living Cells *N. D. Botkin, K.- H. Hoffmann and V. L. Turova*	**109**
Index		**177**

PREFACE

Cryopreservation is a process where cells or whole tissues are preserved by cooling to sub-zero temperatures, typically 77 K. At these cold temperatures, any biological activity, including the biochemical reactions that would cause cell death, is effectively stopped. In this book, the authors discuss the technologies, applications and risks/outcomes of cryopreservation. Topics include cryopreservation of filamentous fungi; differential scanning calorimetry as a useful tool for understanding crystallization and slow-freezing protocols; alternatives to cryopreservation for the short and long-term storage of mammalian cells; and mathematical modeling and simulations in the cryopreservation of living cells.

Efficient mycological (and generally microbiological) work requires a reliable source of cultures, i.e. well-defined and taxonomically determined starting material, which is ensured by its safe long-term storage. Repeated subculturing, frequently used as a routine method for preservation of fungi, is not very practical for storing of large numbers of cultures. It is time-consuming, prone to contamination and does not prevent genetic and physiological changes during long-term maintenance. Various storage methods have been developed in order to eliminate these disadvantages. Besides lyophilization, cryopreservation at low temperatures seems to be a very efficient way to attain this goal. Especially the storage in liquid nitrogen has been considered as the best and most widely applicable preservation technique available for fungi, which seems to surpass all others in the ability to preserve genomic and phenotypic features. It is a safe and reliable method for a long-term maintenance of most fungal species, especially those not amenable to freeze-drying. Agar blocks immersed in an appropriate cryoprotectant were originally used as carriers of fungal mycelium for cryopreservation process, then a useful straw technique with agar miniblocks for the preservation of fungi in liquid nitrogen was developed and subsequently modified by other authors. The cryopreservation process includes freezing and thawing and the protocol of these procedures plays an important part. Different fungal species and even strains can require different cryopreservation protocols. Recently, some new methods of cryopreservation using different carriers of fungal mycelium (perlite, polystyrene or ceramic beads, modified soil, sand, plant seeds etc.) have been developed and successfully verified on a large number of fungal strains from a broad spectrum of taxonomical groups of fungi. Besides survival, another principal requirement for successful maintenance of fungal strains is the ability to preserve their features (growth, morphology, production of metabolites etc.) unchanged. Unfortunately, papers studying this subject are relatively rare, especially in the case of genetic characteristics. It is necessary to add more information concerning the effects of long time maintenance of fungal cultures on their genetic stability using molecular genetic methods. An ideal method should enable preservation of all cultures of fungi keeping their specific characteristics; it should protect them from contamination and at the same time this method should not be expensive and time-consuming. Nevertheless, such universal method has not been created up to now. Therefore it is highly desirable to develop new preservation methods or to improve the current ones, combining advantages and eliminating disadvantages of individual techniques. Besides cryopreservation, also other routinely used preservation methods important for the preservation of filamentous fungi are briefly described in Chapter 1.

In Chapter 2, the control of ice formation and suitable cooling rates are fundamental conditions for the success of slow-freezing protocols. However, only biological studies allow usually evaluation of such protocols, which makes the results dependent on factors external to the congelation.

Consequently, fundamental analytical methods, especially differential scanning calorimetry, have improved our understanding of cryopreservation protocols. Differential scanning calorimetry allows thermodynamic evaluation of cryopreservation solutions under dynamic conditions. Consequently, thermodynamic characteristics, such as the crystallisation temperature, amount of ice crystallised and melting temperature, can be evaluated. In this chapter, after explaining the principle of a differential scanning calorimeter, we will review the main differential scanning calorimetry studies to better understand slow-freezing protocols and crystallisation. In particular, the authors will focus on the characterisation of slow-freezing solutions (part II), optimisation of slow-freezing protocols (part III), elaboration of phase diagrams of solutions containing cryoprotective agents (part IV) and characterisation of intracellular crystallisation (part V).

The ability to preserve mammalian cells has long been a critical part of cell-based research for several reasons. It allows for the transport of cells between laboratories or sites, ensures the availability of consistent starting material for the research through the establishment of cell banks and can uncouple cell-based assays from the culture process. Although cryopreservation, whether conventional slow freezing or vitrification, is widely used as a method for preserving cells long-term, it can result in low cell recovery post-thaw and the cryoprotective agents used in the freezing medium can be cytotoxic. With the emergence of a cell-based therapies industry, where clinical grade cells will need to be stored and transported, there is a growing need to develop scalable GMP-compatible preservation methods that retain not just high cell viability but also clinical efficacy. With these issues in mind, Chapter 3 will explore two alternatives to cryopreservation: cell desiccation for long-term storage of cells and the short-term storage of cells under hypothermic (>0°C) conditions.

Cryopreservation of living cells is a necessary part of many medical procedures such as organ and tissue transplantation, conservation of reproductive and stem cells, etc. However, cells and tissues can be damaged during their freezing and subsequent thawing. Because of complexity of physical and chemical processes, mathematical models are necessary for finding optimal control regimes which increase the survival rate of cells. In Chapter 4, a phenomenological mathematical model based on mean values of physical quantities and a distributed phase field model of ice formation are described, and new methods of their investigation are developed. The results are applied to the design and implementation of optimized cooling protocols in a real freezing plant. The objective of modeling is to reduce cell damage caused, e.g. by sudden temperature drop due to supercooling, by dehydration and shrinkage due to the osmotic flow through cell membranes occurring at slow cooling rates, or by the growth of dendrite seeds that can cause formation of large ice crystals during subsequent thawing. A dynamic programming approach is used as a tool for computing optimized cooling regimes. These techniques are based on effective numerical methods for solving related Hamilton-Jacobi equations. Numerous numerical simulations aimed to the optimization of cooling and thawing protocols accounting for supercooling effects, dendritic growth of ice crystals, and cell shape changes due to osmotic effects are presented.

In: Cryopreservation
Editors: A. Colvert and H. Coty

ISBN: 978-1-62618-474-9

Chapter 1

CRYOPRESERVATION OF FILAMENTOUS FUNGI: A REVIEW

Ladislav Homolka
Institute of Microbiology of the Academy of Sciences
of the Czech Republic, Prague, Czech Republic

Abstract

Efficient mycological (and generally microbiological) work requires a reliable source of cultures, i.e. well-defined and taxonomically determined starting material, which is ensured by its safe long-term storage. Repeated subculturing, frequently used as a routine method for preservation of fungi, is not very practical for storing of large numbers of cultures. It is time-consuming, prone to contamination and does not prevent genetic and physiological changes during long-term maintenance. Various storage methods have been developed in order to eliminate these disadvantages. Besides lyophilization, cryopreservation at low temperatures seems to be a very efficient way to attain this goal. Especially the storage in liquid nitrogen has been considered as the best and most widely applicable preservation technique available for fungi, which seems to surpass all others in the ability to preserve genomic and phenotypic features. It is a safe and reliable method for a long-term maintenance of most fungal species, especially those not amenable to freeze-drying. Agar blocks immersed in an appropriate cryoprotectant were originally used as carriers of fungal mycelium for cryopreservation process, then a useful straw technique with agar miniblocks for the preservation of fungi in liquid nitrogen was developed and subsequently modified by other authors. The cryopreservation process includes freezing and thawing and the protocol of these procedures plays an important part. Different fungal species and even strains can require different cryopreservation protocols.

Recently, some new methods of cryopreservation using different carriers of fungal mycelium (perlite, polystyrene or ceramic beads, modified soil, sand, plant seeds etc.) have been developed and successfully verified on a large number of fungal strains from a broad spectrum of taxonomical groups of fungi. Besides survival, another principal requirement for successful maintenance of fungal strains is the ability to preserve their features (growth, morphology, production of metabolites etc.) unchanged. Unfortunately, papers studying this subject are relatively rare, especially in the case of genetic characteristics. It is necessary to add more information concerning the effects of long time maintenance of fungal cultures on their genetic stability using molecular genetic methods. An ideal method should enable preservation of all cultures of fungi keeping their specific characteristics; it should protect them from contamination and at the same time this method should not be expensive and time-consuming. Nevertheless, such universal method has not been created up to now. Therefore it is highly desirable to develop new preservation methods or to improve the current ones, combining advantages and eliminating disadvantages of individual techniques. Besides cryopreservation, also other routinely used preservation methods important for the preservation of filamentous fungi are briefly described in this review.

Introduction

Fungi represent a significant part of natural ecosystems. They substantially participate in the nutrient cycle and influence the state of health of these ecosystems. Besides the participation in organic compound degradation, numerous fungi play a non-substitutable role in the processes of symbiosis (mycorrhiza, lichenism, symbiosis with animals) and parasitism.

Moreover, many fungi have been used extensively in industrial processes, in basic or applied research, teaching, systematics and biodiversity studies.

Efficient mycological (and generally microbiological) work requires a reliable source of cultures, i.e. well-defined and taxonomically determined starting material, which is ensured by its safe long-term storage. This implies the fundamental and growing importance of culture collections not only for preservation of the endangered genofond (and consequently the biodiversity), but also as a principal source of material for the above mentioned processes. There are almost 500 collections world-wide registered with the World Data Center for Microorganisms (WDCM). However, only 72,000 strains of the estimated 1.5 million fungi have been described and less than 20 % of these are held in collections. It is also evident that many microorganisms have only been isolated once as almost 50% of the species held by the WDCM collections are represented by one isolate only [1]. Conservation of morphological, physiological, genetic, and metabolic stability is crucial for many purposes and is vital for isolates used as medical reference strains, for chemotaxonomic studies, or in the commercial production of biochemicals.

Various storage methods have been developed in order to meet the requirements of the safe maintenance of such material. Among them, the storage in liquid nitrogen (LN) has been considered the best and most widely applicable preservation technique available for filamentous fungi [1]. It is a safe and perspective method of a long-term maintenance of most fungal species, especially those not amenable to freeze-drying. This storage technique seems to surpass all others in the ability to preserve genomic and phenotypic features and is therefore effective in most microbial biodiversity maintenance programmes [2, 3, 4]. But neither this cryopreservation method is applicable to preservation of all fungal cultures in the present form. An ideal method should enable preservation of all cultures of fungi with keeping their specific characteristics; it should protect them from contamination and, at the same, time not be expensive and time-consuming. Nevertheless, such universal method has not been created up to now. Therefore it is highly desirable to develop new or improve the current preservation methods, combining advantages and eliminating disadvantages of individual techniques.

Fungi are a very diverse group and therefore many different methods of cultivation and preservation have been developed to ensure the viability and morphological, physiological, and genetic integrity of the cultures over time. The cost and accessibility of each method must also be considered. Ryan et al. [5] devised a decision-based key, which facilitates the choice of the most appropriate method for long-term preservation of cultures using questions related to fungal characters and user facilities and economics. In order to achieve best practice in the maintenance and provision of biological materials for industry, research and education, the appropriate standards must be followed. The Organisation for Economic Development and Cooperation (OECD) published the guidelines describing how the reproducibility and quality of the technique is maintained in order to retain the full potential of fungi [6, 7, 8]. At present, there is a tendency to concentrate the important biological resources of preserved microorganisms for industry, agriculture and research in Biological Resource Centres, which can offer reliable and reproducible sources of particular cultures [9, 10].

As mentioned above, the increasing biotechnological application of fungi requires, besides their survival, the ability to preserve their properties (growth, morphology, physiological and genetic features etc.) unchanged. Especially for the successful maintenance of production strains, a principal requirement is the full preservation of their important properties (e.g. production of antibiotics, enzymes, and other metabolites). The products resulting after cryopreservation should be checked at morphological (macro- and microscopic features), physiological (growth rate and enzymatic activities) and genetic level (PCR and sequence analysis) to estimate the success of the process. Routine checking carried out in most fungal collections to estimate the success of surviving the process of cryopreservation involve at present usually growth tests together with morphology assessment of fungal colonies, sometimes also certain enzyme production tests. Nevertheless, data on the production of metabolites or other features before and after cryopreservation are relatively rare [11]. The effect of cryopreservation on the production of the antibiotic mucidin by the basidiomycete *Oudemansiella mucida* was studied by Homolka [12], who found that the production ability was practically unaffected. In experiments with some white-rot basidiomycetes, no negative effect of cryopreservation or the cryoprotective used on the production of ligninolytic enzymes was found [13]. Homolka et al. [14] studied production of ligninolytic enzymes laccase and manganese peroxidase and of hydrogen peroxide before and after storage in LN by quantitative spectrophotometric methods, while other enzymes of their set of fungi were estimated by the API-ZYM method. No substantial differences were found. Ryan et al. [15] assessed the effects of preservation regime on secondary metabolite production in two genera of economically important fungi, *Metarhizium anisopliae* and *Fusarium oxysporum*, using thin layer chromatography and high performance liquid chromatography over a 2-year testing period. They investigated five preservation regimes, commonly used in microbial culture collections throughout the world: continual subculture, lyophilization, storage of mycelial plugs in water, storage at -20°C and cryopreservation with liquid nitrogen. Preservation regime influenced secondary metabolite production in the test fungi. Changes in secondary metabolite profiles occurred after relatively short storage periods in most strains, irrespective of the preservation treatment used. On the contrary, both lyophilized and non-lyophilized mycelia of individual isolates of some mycorrhizal fungi showed identical response for all the enzymes tested (nitrate reductase, protease, pectinase, and nuclease). Also physical and functional integrity of lyophilized vegetative mycelium of some tested ectomycorrhizal fungi was comparable to that of non-lyophilized cultures. Inter- and intraspecific variations in morphology, physiology, and metabolic rate were maintained after lyophilization. Maintenance of total protein content confirmed metabolic stability. According to the assays of viability, a plating assay and determination of total biomass confirmed stable mitotic activity of the freeze-dried cultures [16]. Although no preservation treatment can be guaranteed to provide total stability of secondary metabolite production, cryopreservation was the best one of the methods tested [17]. Nevertheless, the survival of cryopreserved fungi is considerably affected by many internal and external factors and different species within a single genus or even strains within a single species can often exhibit different cryoresistance under similar conditions [17, 18, Homolka, unpublished results]. Also the growth and many other requirements of fungal cultures can vary from strain to strain. Therefore, there should be developed new or modified preservation protocols with an emphasis on strain-specific criteria in order to reduce the prospects of instability in physiological features and in secondary metabolite production.

Several interesting studies of genetic stability of the cryopreserved fungi have been performed, but their quantity is still not sufficient. Singh et al. [19] confirmed genetic stability of 11 cryogenically preserved edible mushroom strains by comparing Random Amplified Polymorphic DNA (RAPD) and Internal Transcribed Spacers (ITS) amplified profiles, Ryan et al. [20] checked genetic stability using polymerase chain reaction (PCR) fingerprinting in several isolates of *Fusarium oxysporum* and *Metarhizium anisopliae* (unfortunately not in their third model *Serpula lacrymans*), Stummer et al. [21] confirmed genetic stability of *Uncinula necator* conidia after storage at -70°C, and Voyron et al. [22] showed in their study of the influence of the mid-term cryopreservation at -70°C on genetic stability of 15 isolates (10 species) of basidiomycetes that DNA fingerprint patterns obtained by the AFLP analysis were not changed. Camelini et al. [23] achieved preservation of *Agaricus subrufescens*, one of the most important fungi produced biotechnologically in Brazil, without significant morphological and genetic changes within the 12-month storage period. They used mycelium on a solid medium supplemented with activated charcoal. Genetic stability was assessed by RAPD analysis. RAPD profiles demonstrated high genetic homogeneity of the mycelia using this technique (> 99%). Also sequence analysis of eight strains of basidiomycetes did not detect any changes in their genetic features after cryopreservation [24]. On the contrary, AFLP analysis in *Cochliobolus sativus* revealed several genetic changes between stored and non-stored isolates. A total of 444 scorable DNA bands were obtained; 324 of these (71%) were polymorphic. However, the proportion of DNA alterations varied according to the storage method [25]. Broughton et al. [26] assessed genomic integrity in five strains of the economically important fungus *Trichoderma*. They applied two PCR fingerprinting techniques and detected amongst replicates three post-preservation genetic polymorphisms, indicating that even standardized cryopreservation methodologies can sometimes induce genomic change. However, the low number of polymorphisms suggests that cryopreservation is a reliable method for organism storage over long periods of time. Especially genomic and proteomic research programmes with fully characterized organisms should be supplied, as usage of organisms from unreliable sources can prove disastrous, not least in economical terms [27].

A lot of reviews on preservation of microorganisms including cryopreservation of filamentous fungi have been published. A very comprehensive and still useful book with detailed overview of the methods and results of cryopreservation of microorganisms, including fungi, was published by Hubálek [11]. Abd-Elsalam et al. [28] discuss the importance of culture collections in plant pathology and review the methods currently available to store cultures. The preservation and maintenance of plant pathogenic fungi in a viable yet stable state for long periods are important, because isolates of these fungi can serve as standards for identification of quarantine taxa. Such isolates are also important for testing disease resistance and for plant breeding programs. The increasing use of molecular sequences analysis in the systematics of plant pathogenic fungi has meant that maintaining fungi in culture collections has become essential. Tedeschi and De Paoli [29] summarize the most important steps and components in the process of low- and ultra-low temperatures freezing of bacteria, parasites, yeasts and fungi, viruses, and recombinant microorganisms. Another review on preservation of microorganisms published Costa and Ferreira [30], of fungi Tan [31]. A very intelligible review of preservation and distribution of fungal cultures with useful protocols for each method was published by Nakasone et al. [32]. It is desirable to add more information on the longer time of culture maintenance (at least 10 years of LN storage) on the preservation of different metabolite production and activities and on the genetic stability of the cryopreserved cultures using molecular genetic methods. Nagai et al. [33] evaluated preservation techniques used for 14,836 microorganisms in the Genebank in Tsukuba, Japan. Many further reviews have been published during previous decades [see 11].

Description of cryopreservation devices, mechanisms of cryoprotective action and cryoinjury or detailed characterization of cryoprotectants are out of the scope of this review and can be found in the specialized literature [e.g. 11, 34, 35, 36, 37, 38, 40, 41, 42, 43, 44, 45]. The same holds for physical principles of cryogenics [46, 47, 48, 49]. The books by Muldrew and McGan [47] and by Katkov [49] are available for free at the Internet.

Many methods have been developed for the preservation of fungi; all of them have their advantages and disadvantages and so the method most suitable for a particular need must be determined considering its important features like maintenance of viability, genetic stability, purity, but also frequency of use of cultures, their number, and the cost of culture maintenance. Basic methods such as continual subculture/serial transfer in agar, storage under mineral or paraffin oil

[50, 51, 52], in sterile water [53, 54, 55, 56], cool storage in a standard refrigerator at 5-8°C, storage in sand, grain or soil [57], or on silica gel, may be suitable in the absence of alternatives. The majority of large collections ensure that each of their strains is preserved by at least two methods. For most, the preservation methods of choice are freeze-drying and cryopreservation in LN.

Except for subculturing, all the methods are based on at least partial suppression of growth and metabolism of the fungus. The cultures can be dried (desiccated) by removal of water and by prevention of rehydration using soil, silica gel, sand, paper strips, and various other materials as carriers. Specific derivations of subculturing are storage of cultures on agar in the refrigerator, under oil or in sterile water. Because the stability of fungal cells was not ensured by these simple procedures, other methods have been suggested, such as cryopreservation either in LN using controlled-rate freezing or at low temperature (-20 and -70°C), and lyophilization (freeze-drying) using a shelf or centrifugal freeze-drier. Cryopreservation in liquid nitrogen and lyophilization are the methods recommended and used by the American Type Culture Collection [58] and other leading world collections. Cultures that are permanently preserved in metabolically inactive states now can serve as type specimens, according to Article 8.4 of the International Code of Botanical Nomenclature [59]. More recently, techniques such as vitrification and encapsulation cryopreservation have been used to preserve recalcitrant fungi.

The primary methods of culture preservation are then continuous growth, drying, and freezing. Continuous growth methods are simple and inexpensive, because specialized equipment is not required. Drying is the most useful method of preservation for cultures that produce spores or other resting structures. Silica gel, glass beads, and soil are substrata commonly used in drying. E.g., fungi have been stored successfully on silica gel for up to 11 years [60]. Also drying methods are technically simple and do not require expensive equipment. Freezing methods, including cryopreservation, are versatile and widely applicable. Most fungi can be preserved, with or without cryoprotectants, in LN or in standard freezers. With freeze-drying (lyophilization), the fungal cultures are frozen and subsequently dried under vacuum. The method is highly successful with cultures that produce spores. Freeze-drying and freezing below -139°C are excellent methods for permanent preservation. However, both methods require specialized and expensive equipment.

Many fungi, esp. basidiomycetes, do not form asexual spores, their dominant form being mycelium which is rather sensitive to environmental conditions, and also mycorrhizal relationships play often a certain role. Recently, some interesting methods for cryopreservation of basidiomycete cultures were reported. One is the

Mata's freezing method, in which wheat seeds of *Sorghum vulgare* were used as a medium for culturing and stocked as a frozen spawn, and the other is Homolka's perlite protocol (PP) using perlite as a carrier of mycelium [61]. This protocol was then successfully verified on a large number (442) of basidiomycete strains [14] and can also be used for cryopreservation of taxonomically different groups of fungi including yeasts [62] and for fungi which failed to survive other routinely used preservation procedures.

Besides cryopreservation, also other routinely used preservation methods important for the preservation of filamentous fungi, are briefly described in this review.

Preservation Methods

The primary methods of culture preservation are continuous growth, drying, and freezing. A traditional method of serial transfers, which is still frequently (sometimes indispensably) used in many laboratories, is not very practical for preservation of large numbers of fungal cultures. It is prone to contamination by other microorganisms or mites, is labour-intensive and space-consuming [63], and does not prevent genetic and physiological changes during long-term maintenance [64]. Various storage methods have been developed in order to eliminate these disadvantages. Among them, besides lyophilization, the storage in liquid nitrogen (LN) has been considered the best and most widely applicable preservation technique available for filamentous fungi, which seems to surpass all others in the ability to preserve genomic and phenotypic features [1]. It is a safe and reliable method of a long-term maintenance of most fungal species, especially those not amenable to freeze-drying. This method is useful for both sporulating and non-sporulating (mycelial) fungal cultures and has several other important advantages: protection of cultures from contamination (reduced number of transfers) and the saving of time, work and room.

The most common method to preserve fungi in culture collections is *continuous subculturing* on agar slants, which is sometimes nicely named micro-gardening [65], has been described in many papers [e.g., 66, 67, 68]. However, this method has a lot of the abovementioned disadvantages, namely it does not prevent damage to the cultures consisting in an unpredictable appearance or selection of diverse phenotypic variants (pleomorphic growth) and genetic mutants. These changes might result in cultures sometimes completely different from the fungal populations originally isolated or acquired. Therefore this routine subculturing is not a very practical for storing large numbers of fungal cultures.

On the other side, this technique is applicable universally and is at present used in a majority of culture collections with limited funds and also in many collections (including the prominent ones) as a second method besides cryopreservation or lyophilization. In many papers, subculturing is compared with other methods of maintenance [e.g., 69, 70 and many others].

Continuous growth methods, in which cultures are grown on agar, are typically used for short-term storage. Such cultures are stored at temperatures of 5-20 °C, or they may be frozen to increase the interval between subcultures. The methods are simple and inexpensive because specialized equipment is not required.

One easy method of reducing frequency of subculturing is by lowering the rate of metabolism by *cold storage*. At its simplest the cultures are stored in a standard refrigerator at 5-8 °C. E.g., working cultures are kept in this way at the Commonwealth Mycological Institute (CMI). The Centraalbureau voor Schimmelcultures (CBS) maintains many isolates by frequent transfer and stores them at 15 °C and 60% relative humidity, which avoids cold damage but still reduces growth rate.

Perlite can be used for the cultivation of fungi by serial transfers [71]. Cultures of 33 basidiomycete strains out of 35 tested were viable with unchanged characteristics after four years of maintenance on perlite in cryovials at 4°C. These cultures can be a good substitute for agar cultures in medium-term maintenance of fungi. For comparison, the storage under mineral oil was also evaluated, but it turned out to be unsuitable for the majority of the Culture Collection of Basidiomycetes (CCBAS) cultures.

The major drawbacks of sequential subculturing can partially be eliminated by several modifications consisting mainly in slowing down the fungal metabolism thus enabling the intervals between the culture transfers to be prolonged. Such adjustments include, for example, maintenance of the grown cultures under sterile mineral oil, in distilled water, exsiccation on different porous substrates (silica gel, sand, glass or porcelain beads, gelatine discs, soil, etc.) or dried from a liquid state (L-drying). A more sophisticated and efficient technique is freeze-drying (lyophilization), i.e. quick drying of frozen fungal suspensions by sublimation in vacuum. A common disadvantage of all these preservation techniques is that fungi can lose their viability or degenerate rather quickly. New storage methods are therefore being developed and tested. For example, the controlled-rate freezing and drying might be useful for some mycelial fungi refractory to standard freeze-drying [72, 73]. A combined technique of maintaining freeze-dried microbial preparations at ultra-low temperatures (in LN vapours, in a cryogenic freezer) has also been described [74, CBS at Utrecht, personal communication].

Preservation under mineral oil is cheap and easy and is recommended for the laboratory with limited resources. Cultures can be kept for several years or, in exceptional cases, up to 32 years at room temperature or 15-20°C. This method is appropriate for mycelial or nonsporulating cultures that are not amenable to freezing or freeze-drying. As an added benefit, oil also reduces mite infestations. Although many fungi can be maintained this way, the growth rates of the cultures may decrease as storage times increase [52, 75]. Onions [76] reported revivals of 21 years for such cultures maintained in the Commonwealth Mycological Institute (CMI) collection. The CBS also stores a part of its collection under mineral oil. Kochetkov [77] studied viability of 105 strains belonging to 47 species after 25-30 year storage under vaseline oil and did not find any correlation between culture viability and systematic position. Piaggio [78] reported quite high viability of 127 fungal strains (mainly wood-rotting basidiomycetes) stored under oil and retrieved after 10 years and 20 years. A general survival rate was 80% after 10 years and 93% of the surviving after 20 years. Except for *Zygomycotina*, the storage of living fungal cultures under mineral oil was relatively successful. Perrin [51] recommend this method for species or isolates that resist freeze-drying such as mycelial cultures of wood-destroying fungi. Nevertheless, the procedure did not prove to be suitable for the maintenance of basidiomycete cultures tested by Homolka and Lisá [71]. The major disadvantage of the maintenance under mineral oil is that the fungi continue to grow, and thus, a selection for mutants that can grow under adverse conditions may occur.

Storage in sterile water is another inexpensive method for maintaining fungal cultures that has aroused recent interest. It was originally described by Castellani [79, 80] who stored fungi pathogenic to man. Ellis [55] obtained good viabilities of up to 20 months of isolates that did not sporulate well or gave poor results with lyophilization, including *Entomophtorales*, *Pyrenomycetes*, *Hymenomycetes*, *Gasteromycetes* and *Hyphomycetes*. Apparently, distilled water suppresses morphological changes in most fungi. The method has been successfully used to preserve oomycetes [e.g. 60, 81], basidiomycetes [e.g. 56, 75, 82] ectomycorrhizal fungi [e.g. 54, 83], ascomycetes [e.g. 52], plant pathogenic fungi [e.g., 84], and human pathogens and yeasts [e.g. 85]. Most basidiomycetes survived for at least 2 years at 5°C [54, 56]; viability decreased after 5-10 years of storage [75]. Although Ellis [55] reported that most of the basidiomycetes he tested survived for 20 months when stored at 25°C, Johnson and Martin [52] recovered only 26% of the basidiomycete strains stored at 20°C. Ascomycetes, however, including their mitosporic forms, survived up to 10 years when stored at 20°C [52]. The disadvantages of this method are the same as in subculturing.

It belongs to the techniques by which it is possible to postpone the requirement to subculture [86]. Boeswinkel [84] could store as many as 650 diverse plant pathogens in water for 7 years without loss in their viability and pathogenicity. It is possible to preserve the viability of some fungi for at least 20 years by this method [87]. Up to now, only spores, hyphae and mycelium on agar blocks have been stored in sterile water [63]. Marx and Daniel [54] stored mycelial agar discs of ectomycorrhizal and plant pathogenic fungi in sterile water with good results. McGinnis et al. [85] used sterile distilled water to store hyphal and spore suspensions of 147 different species of fungi at 25°C for periods of 12 to 60 months. These included more than 25 species of dermatophytes. Storage in sterile distilled water or lyophilization of dermatophytes was successful with high survival rates (89.7% or 87.2%) after 12 years of preservation of 78 isolates belonging to seven genera [88].

Apart from losing virulence and pathogenicity, fungi may also lose some of their characteristic morphological and physiological features when repeatedly subcultured [89] or even during cryogenic storage [90]. For instance, the synnematous varieties of *Hirsutella thompsonii* Fisher (*Ophiocordycipitaceae*), lose their ability to produce synnemata especially when the basal mycelium is used for subculturing over and over again [91]. On the contrary, Kumar and Singh [92] successfully stored five nonsynnematous and six synnematous isolates of the acaropathogenic fungus *H. thompsonii* as mycelial beads in sterile water under refrigeration. They were viable for up to 96 months without any adverse effect on their pleomorphic characteristics, pathogenicity and genetic purity.

A typical example of the advantages and disadvantages of preservation of fungi in water was provided by Borman *et al.* [93]. They evaluated the survival and potential morphological alterations of 45 species of pathogenic filamentous fungi that had been stored in sterile water following Castellani's method in the National Collection of Pathogenic Fungi (NCPF). Storage duration varied from 2 months to over 21 years. 90% of stored organisms were shown to be viable. Viability was largely independent of the duration of storage, but did apparently vary to some degree in an organism-specific manner. In addition, certain fungi were shown to have undergone morphological alterations during storage, and exhibited significant degrees of pleomorphism upon revival. This was especially marked for several isolates of dermatophytes, where storage resulted in loss of recognisable colonial features, and overproduction of sterile mycelium with aberrant or no conidia. These findings suggest that while Castellani's method remains a useful method for long-term preservation of many fungi, water storage should be supplemented by a second storage method to increase the chances of retaining both viability and morphological stability over long periods.

Storage under sterile water is also a suitable method to store isolates of basidiomycete decay fungi for long periods (18 years), which was confirmed by the study of Richter et al. [94]. In her previous study [95], she published evaluation of 69 basidiomycete isolates grown on vegetatively colonized agar (48 species in 30 genera and 17 families) and stored at +5°C in tubes of sterile distilled water without manipulation for 20 years. These were represented by 34 isolates of saprotrophic fungi (29 species in 19 genera) and 35 isolates of mycorrhizal fungi (19 species in 11 genera). Viability evaluation was based on reviving growth on agar media at room temperature. 57 out of 69 isolates (82.6%) grew vigorously when revived after storage for 20 years (88.2% of the saprotrophic isolates, 77.1% of mycorrhizal. 13 isolates of *Laccaria* were all viable after 20 years, indicating cold storage in sterile water to be a good method for maintaining this important genus of mycorrhizal fungi.

Desmukh [96] successfully used simple methods (maintenance on agar slopes in a refrigerator at +8°C and in sterile distilled water, under liquid paraffin, dried on silica gel, in sterile soil at room temperature) for preservation of a total of 239 fungi belonging to 9 genera and 23 species of keratinophilic fungi and related dermatophytes.

Drying enables suspending of fungal metabolism to a level at which metabolism is either reduced or preferably stopped but from which the culture can be revived. This method of preservation is useful for cultures that produce spores or other resting structures. Its disadvantages are culture degeneration if moisture or strain drift [32]. Silica gel, glass beads, and soil are substrates commonly used in drying. Fungi have been stored successfully on silica gel for up to 11 years [60]. Drying methods are suitable for laboratories with limited funds because they are technically simple and also do not require expensive equipment. These laboratories can use the silica gel technique for preserving sporulating fungi of the *Ascomycota* and many species of mitotic fungi. Using this method, 64% of 421 strains survived the storage for a quarter of a century or more [97]. Nevertheless, drying technologies tested for preservation of aerial conidia of *Beauveria brongniartii* and *Metarhizium anisopliae* resulted in severe damage to the fungal material [98]. Quintanilla et al. [99] determined spore germination of five globally threatened fern species (*Culcita macrocarpa*, *Dryopteris aemula*, *D. corleyi*, *D. guanchica* and *Woodwardia radicans*) after 1, 6 or 12 months of storage in glass vials (dry storage) or on agar (wet storage) at -20, 5 or 20°C. In all species, storage technique, storage temperature and the technique-temperature interaction all had a significant effect on germination percentage. In most cases, the germination percentage was best maintained by wet storage at 5 or 20°C. Wet storage at -20°C killed all or most spores in all species. Dry storage, which has lower preparation time and space requirements than wet storage, was generally more effective at the lower temperatures (-20 or 5°C).

Another possibility is to use *organic substrates* [32]. Many different practical methods of preserving fungi on various organic substrates (e.g., wood chips, cereal grains, straw, filter paper, insect or plant tissues) were developed. For instance, some wood-inhabiting basidiomycetes and ascomycetes can be stored on wood chips for up to 10 years [100]. Species of *Sclerotinia, Magnaporthe, Leptosphaeria* and *Rhizoctonia* species have been stored for up to 10 years on seeds of oats, barley, wheat, rye, millet, and sorghum [101, 102]. Nuzum [103] described a method of vacuum-drying fungal cultures on agar strips. *Pythium, Rhizoctonia* and some basidiomycete species survived 18 months with this method, whereas ascomycetes and their mitosporic forms survived for 3-5 years. Pathogenic fungi can be stored on their host tissues, e.g. *Pyrenochaeta* and *Thielaviopsis* on plant roots [102] or conidia of *Neozygites fresenii*, which cannot be cultured in vitro, on aphid mummies [104].

Sterile soil or sand was used for a long-time successful preservation of *Rhizoctonia* [101], *Septoria* [105], and *Pseudocercosporella* [106]; nevertheless, morphological changes in some fungi have been recorded. Storage of spores of vesicular-arbuscular mycorrhizal fungi in soil at 5°C is a common way of preserving these fungi.

Perkins [107] developed the *silica gel* method originally for *Neurospora* species. He found that sporulating fungi protected by skim milk and stored on silica gel remain viable for 4-5 years, while Raper [108] preserved spores and microcysts of dictyostelids for up to 11 years. The method can be used to preserve sporulating fungi if facilities for freeze-drying or for storage in liquid nitrogen are not available. Fungi such as *Pythium* and *Phytophthora* species, however, do not survive this process. Some researchers use *glass beads* instead of silica gel [109].

Freeze-drying, another term for *lyophilization*, implies the common features of this process with cryopreservation. Dehydration of cells, necessary to avoid intra-cellular crystallization during freezing, is regulated through the cooling rate and depends on the size of the cell and the thickness of the cell-wall. Saccharides prevent phase transitions in the membrane from the liquid-crystalline to the gel-phase during dehydration. Proteins are protected against denaturation during slow cooling by saccharides and amino-acids. During freeze-drying, cells are dehydrated in the cooling step preceding drying; during drying the protectant is converted in a glass. Cryoprotectants such as glycerol, dimethyl sulphoxide, 1.2 propane diol, ethylene glycol, ethanol, methanol, polyethylene glycol and several butane diols serve to avoid extra- and intracellular crystallization, but most of them are toxic. Revival in 1.2 M sucrose is recommended [110]. Interesting view of freeze-drying process was published by Franks [111].

The centrifugal dryer exploits the fact that drying under reduced pressure is accompanied by the drop in the temperature. Water is then removed by the sublimation of ice from the frozen sample [32]. Drying must avoid the liquid phase and be executed at temperatures below -15°C until the water content of the culture is reduced to less than 5% [43]. Another lyophilization method uses a standard shelf freeze-drying and skimmed milk as a suspending solution/lyoprotectant. This method has been employed to freeze-dry the majority of the 50,000 fungal strains that have been successfully lyophilized at the CBS culture collection [112]. Besides cryopreservation, lyophilization is the most popular method of fungus culture preservation. It was first applied to fungi in 1942 at the Northern Regional Research Centre at Peoria, Illinois [113]. It is so successful for the majority of spore-bearing fungi that it has been adopted by most major culture collections, which report at intervals ever lengthening viabilities. Lyophilized spores of dictyostelids, with associated bacteria, were maintained successfully for up to 30 years [108]. Freeze-drying is the preservation method of choice for many spore-forming fungi that produce large numbers of spores 10-μm or less in diameter. Larger spores tend to collapse during the lyophilization process, and the structural damage caused is not reversible by hydration. A significant number of the spores of appropriate size are also physically damaged and killed during the freezing process by the formation of ice crystals. Rapid freezing and the addition of a lyoprotectant minimize the growth of ice crystals. Bunse and Steigleder [114] presented a simplified method for the preservation of fungal strains by lyophilization. The cultures are grown on Sabouraud glucose agar in glass ampoules and lyophilized without further processing. By this method the macroscopical morphology of the cultures is preserved, so that these can be used immediately without recultivation as reference cultures. All tested mould and yeast strains remained viable over the six months duration of the experiment, whereas the dermatophyte strains could only partially be recultivated. Good results were obtained with *Penicillium*, *Aspergillus*, *Cladosporium*, *Alternaria*, *Mucor*, *Candida*, and *Rhodotorula* species. Suspending media (media with lyoprotectants) can protect from freezing damage, dehydration, oxygen damage and storage problems. Freeze-dried ampoules are usually sealed under vacuum, because oxygen in the system will cause loss of viability. Most organisms survive well at room temperature, but it is considered preferable to store at 4°C. At CBS in Utrecht the lyophilized cultures are stored at ultra-low temperatures. While in sporulating cultures lyophilization is a method of choice, in non-sporulating ones the situation is more complicated and mycorrhizal cultures are most problematic. However, Croan [115] demonstrated that mycelial isolates of non-sporulating basidiomycetes can be lyophilized effectively in the presence of trehalose. The effort to apply lyophilization to fungal hyphae in most cases has met with mixed success [72, 73], although *Claviceps* spp. [116] are reported to have survived. Useful lyophilization protocols published also Tan et al. [112] and Ryan and Smith [117]. Sundari and Adholeya [118] developed a successful lyophilization procedure for the basidiomycete *Laccaria fraterna*, a filamentous non-sporulating (*in vitro*) fungus, by optimization of the freeze-drying protocol. The culture retained its viability after lyophilization and when subjected to quality assurance tests, it gave consistent results similar to the non-lyophilized culture, indicating the stability of the product and the applicability of the developed process to freeze-dry vegetative mycelium of filamentous non-sporulating fungi. Even the enzyme activity was retained following freeze-drying the mycelium of ectomycorrhizal isolates [119]. Lyoinjury, damage caused by the freeze-drying process, can occur during the cooling and/or drying stages. The phase changes encountered during the drying process can cause the liquid crystalline structure of the cell membranes to degenerate to the gel phase, which disrupts the fluid-mosaic structure of the membrane [110]. Lyoprotectants such as serum, skimmed milk, inositol, trehalose and peptone can be added to the suspending medium to reduce the prospects of damage. Once preserved, cultures are easy to handle and occupy relatively little storage space with a proven shelf life of 20-40 years or longer [63].

Croan [115] investigated alternative methods of lyophilization of mycelial isolates of tropical wood-inhabiting basidiomycetes which are not normally considered candidates for preservation by lyophilization, because these fungi do not form resistant propagules. The tested fungi survived lyophilization without noticeable changes in viability, growth rates, or morphological and biochemical characteristics. In her previous paper, Croan with colleagues [82] compared storage in water, lyophilization, and cryopreservation methods of preserving tropical wood-inhabiting basidiomycetes. When mycelial isolates were stored in sterile distilled water, 57% survived at 4°C for 2.5 months; 92% of the isolates survived at 15°C for 10 months. Bulbil-, clamydospore-, and conidium-forming fungi survived in cold water storage and maintained excellent viability for a decade. Vegetative hypha-, bulbil-, clamydospore-, and conidium-forming fungi survived cryopreservation using liquid nitrogen vapor. Only basidiomycetes that formed bulbils, chlamydospores, or conidia survived lyophilization.

Viability and *in vitro* and *in vivo* activities of freeze-dried conidia of the predatory fungus *Arthrobotrys robusta* were evaluated after a month storage at 4°C against infective larvae of *Ancylostoma* spp. in dogs. These conidia survived freeze-drying, the passage through the gastrointestinal tract of the treated dog and were able to grow in the plates, form traps and capture the larvae [120]. Toegel et al. [121] devised a modified method of lyophilization of aerial conidia and biomasses from entomopathogenic fungi *Beauveria brongniartii* and *Metarhizium anisopliae* using combination of fructose and dextran 4. In comparison to freshly harvested biomasses, the lyophilized products showed similar growth rates and comparable production of virulent secondary metabolites such as destruxin A, destruxin B, or oosporein, suggesting their applicability as biological control agents. Not very impressive results were obtained by Rybnikář [122] in 26 sporulating strains of lyophilized microscopic fungi of the genera *Epidermophyton, Microsporum, Paecilomyces* and *Trichophyton* maintained at 2-8°C for 8-14 years. In 23 *Trichophyton* strains the number of viable particles was reduced almost 10-fold during this period. The viability of other three investigated strains ranged from 3.4% to 7.1% in comparison to the original values. The recovery of 75 filamentous fungal strains (*Deuteromycotina*, *Ascomycotina* and *Zygomycotina*) and 10 yeast cultures which have been preserved for 49 years in a lyophilized state examined Schoenleincrusius et al. [123] with concentration on morphological and taxonomical features to check the identification. From the total 85 lyophilized isolates, 55% (43 filamentous fungal strains and 6 yeast cultures) was recovered.

Storage at low and ultra-low temperatures (in freezers and liquid nitrogen), i.e. cryopreservation in the real sense of the word is at present the most important technique used in preservation of fungi. A comprehensive and detailed overview of the methods and results of cryopreservation of microorganisms was published by Hubálek [11]. Another very useful and intelligible review was published by Nakasone et al. [32]. As cryopreservation is at present extraordinarily widespread technique, a great number of reports on this topic have been published [e.g., 1, 124 and many others].

Cryopreservation is versatile and applicable to a wide spectrum of fungi including those that are not amenable to lyophilization. For instance, many non-sporulating or fastidious mycelial fungi (*Pythium*, *Achlya*, *Entomophthorales*, *Cunninghamella*, *Basidiomycota*) or fungi producing large and/or thick-walled conidia cannot be lyophilized successfully [63, 125]. Different fungal species and even strains can require different cryopreservation protocols. Recently, some new methods of cryopreservation using different carriers of fungal mycelium (perlite, polystyrene or ceramic beads, modified soil, sand, wheat seeds etc.) have been developed and successfully verified on a large number of fungal strains from a broad spectrum of taxonomical groups of fungi. Last but not least, cryopreservation at ultra-low temperatures is the method of choice for genetically manipulated microorganisms.

Most fungi can be preserved, with or without cryoprotectants, at -20° to -80°C in *mechanical freezers* and remain viable. Cultures grown on agar slants in bottles or test tubes with screw caps can be placed directly in the freezer [126]. Overall failure rate for mitosporic ascomycetes, zygomycetes, and yeasts after 5 years in storage at -20°C was 5.1% [127]. The failure rate of medically important fungi, aerobic actinomycetes, and algae stored from 6 months to 13 years at -70°C was 2.3% [128]. Ito and Yokoyama [129] aseptically removed six 6-mm discs from vigorously growing cultures of nonsporulating basidiomycetes and ascomycetes and placed them in sterile cryotubes containing 10% glycerol in water. The cultures were preserved successfully for up to 5 years by mechanical freezing at -80°C. The cultures were then checked again several times [130, 131, 132]. Fungi grown on various organic substrates, such as cereal grains, agar strips, plant parts, and filter paper, and then dried can also be frozen. In general, vigorously growing and sporulating cultures survive the freezing process better than less vigorous strains. Repeated freezing and thawing significantly reduce viability of the cultures [32]. Carmichael [126, 127] recommended the storage at -20°C for sensitive medical fungi. Agar slope cultures can be stored successfully in a freezer at about -20°C. Camelini et al. [23] achieved preservation of *Agaricus subrufescens*, one of the most important fungi produced biotechnologically in Brazil, without significant morphological and genetic changes within the 12-month storage period. They used mycelium on a solid medium supplemented with activated charcoal. Genetic stability was assessed by random amplified polymorphic DNA (RAPD) analysis. RAPD profiles demonstrated high genetic homogeneity of the mycelia using this technique (> 99%). Commercial preservation systems with polystyrene beads as carriers were used for cyopreservation of conidia of enthomopathogenic fungi [133] and of sporulating *Aspergillus fumigatus* cultures at -80 °C [134]. Cryopreservative-immersed porous ceramic beads were employed for cryopreservation of several sporulating fungal cultures of biotechnological importance (*Aspergillus*, *Mucor*, *Phaffia*, *Rhizomucor* and *Saccharomyces*) at -70°C by Palágyi et al. [135]. Polyethylene glycol added after the transfer of cells into the vial increased the safety of storage. It is symptomatic that these techniques have not been used for non-sporulating filamentous fungi. In this context it should be mentioned that as early as 1978 Feltham et al. described a method of preservation of bacteria on glass beads at -76°C. In general, the storage of cultures at the temperature between -15 and -40°C is not suitable for the majority of them. Also the fluctuation of temperature during the storage affects negatively their viability.

Storage in *liquid nitrogen (LN)* has been considered the best and most widely applicable preservation technique available for filamentous fungi [1]. It is a safe and reliable method of a long-term maintenance of most fungal species. LN refrigeration has, of course, also some disadvantages, e.g. the necessity of LN supply (the need to continuously replace LN can be difficult at small or remote facilities), possible cryovessel failure, initial investments into the equipment, LN contamination or problems with culture distribution (strains preserved in LN must first be grown on agar or in liquid medium to avoid the expense of shipping frozen materials), but the advantages exceed the drawbacks substantially. LN refrigeration permits a long-term (theoretically infinite) survival. The only limitation to the period of storage in LN might be the accumulation of damage caused by background radiation to which the stored microorganisms are accessible [34, 136].

Different protocols for cryopreservation using LN have been published [e.g., 4, 60, 137, 138, 139, 140, 141, 142 and many more], varying with equipment, special needs, preference of materials or the type of organism.

In the pioneering work of Hwang [143, 144], agar blocks immersed in a cryoprotectant were used as carriers of fungal mycelium for the cryopreservation process. Mycelium and/or spore suspensions with or without a cryoprotectant in sealed borosilicate glass ampoules were originally used for cryopreservation of filamentous fungi [145]. These ampoules are not safe because LN can penetrate into improperly sealed ampoules which are likely to explode when removed from the storage vessel due to the rapid expansion of contained nitrogen gas [146]. Later, glass ampoules were replaced with safer polypropylene cryovials which are also more easily manipulated and stored. Their use even improved viability of several strains of fungi (*Basidiobolus*, *Rhizophidium*, *Phytophtora* and *Serpula*) in comparison with glass ampoules [147]. Several alternative forms of plastic ampoules have then been developed. Simione et al. [148] used plastic ampoules with screw caps which would not explode even if penetrated by nitrogen. A number of manufacturers (e.g., Nunc, Nalgene, Sterilin, Greiner, Vangard) produce autoclavable polypropylene and polycarbonate cryovials (cryotubes) with internal or external threads and ring seals, which are nowadays routinely used in many collections and laboratories. The useful straw technique for the preservation of fungi in LN using agar miniblocks in polypropylene (drinking) straws was developed by Elliott [139] and improved by Stalpers et al. [140]. Since then the straws have been often used in cryobiology [149, 150, 151]. In Stalper´s procedure the straws are sealed at one end, filled by agar miniblocks or spore suspension using a long hypodermic needle, sealed at the other end, and tested for leaks by pressing them against a hard surface. This procedure was modified by other authors [152, 153] who used straws sealed on one or both ends or unsealed at all and closed in cryotubes. Thammavongs et al. [154] evaluated the technological potential of ionomeric resin straws compared to polycarbonate cryotubes. Survival under freezing stress was tested on three microorganisms selected for their biotechnological interest: two lactic acid bacteria, *Lactococcus lactis* subsp. cremoris and *Lactobacillus delbrueckii* subsp. bulgaricus, and a deuteromycete fungus, *Geotrichum candidum*. The stress was carried out by repeated freezing-thawing cycles to artificially accelerate the lethal effect of freezing on the microorganisms. Conservation in the resin straws led to a slight increase in the survival of *L. cremoris* and *G. candidum* compared to microtubes. Homolka et al. [153] compared viability of basidiomycete strains after cryopreservation by two different freezing protocols - the first one according to Hwang [155] and the second one using a modified protocol of Hoffmann [152].

In case of the original agar plug protocol, 162 strains (65%) of the 250 strains survived a 2-day storage and 158 strains (63 %) survived a 2-year storage in LN. In case of the straw protocol, 246 strains (98 %) of the 250 strains survived a 2-d storage and 243 strains (97 %) a 2-year storage in LN. The results indicate that the protocol used for cryopreservation can significantly influence strain survival. Markedly better results were obtained using the straw protocol. Tuite [156] sealed fungal cultures in a polyester film and froze them in LN. However, the frozen film is very fragile and not very suitable for routine mass work.

Porous beads represent an alternative, handy method of maintaining fungal suspensions at low temperatures. The method was designed originally for storing bacteria in a deep freeze at -70°C [157], but it can be used for storing fungal spores [133] and other microorganisms in the vapour above LN. Microbial cells adhere to small (cca 3 mm diam.) beads placed in a 2-ml cryovial; 20-30 beads of a strain can be stored in one cryotube. Chandler [133] described a simple method for the cryopreservation of conidia; they were stored being adhered to small porous beads in a robust polypropylene vial. For recovery, a single bead is removed from the vial and streaked onto a solid growth medium. Preparative work is minimised so the method is rapid.

A new method of cryopreservation using perlite as a carrier of fungal mycelium (perlite protocol, PP) was developed by Homolka et al. [61] originally for 5 basidiomycete strains and then successfully verified on 442 basidiomycete strains [14]. Perlite is a unique aluminosilicate volcanic mineral holding and retaining substantial amounts of water, which can be released as needed. Expanded perlite has been used in many applications, particularly in the construction, horticulture, and other various industrial fields. Before the cultures are placed into LN or its vapours, or in a cryogenic freezer, they are usually pre-cooled in a controlled-rate cooler (programmable freezer) or another alternative device, e.g. a deep-freezer (-65°C), dry-ice box, or polystyrene box. A very simple method for adopting a slow, cca 1°C/minute cooling rate is to use a small commercial ´Mr Frosty´ container filled with isopropyl alcohol. The cryopreserved cultures are kept in different types of cryovessels and cryocontainers with LN. An alternative to LN are cryogenic electro-mechanical freezers, which are able to keep the temperature between -140 and -150°C, i.e. below the glass transition temperature of ice. CBS in Utrecht is equipped with ultra-low temperature freezers placed in tanks, which can be filled with LN if needed [Stalpers, personal communication]. Because the rates of mutation in cultured fungi are likely to correspond to those of cell division and metabolic activity, storage methods that stop cell division completely and totally arrest metabolism, while still retaining viability, are best. Freezing at or below -139°C, the temperature at which ice crystals do not grow and rates of other biophysical processes are too slow to affect cell survival, accomplishes this.

Deutsche Sammlung von Mikroorganismen und Zellkulturen GmbH (DSMZ), CBS and The Belgian Co-ordinated Collections of Microorganisms (BCCM) prepared in 1998 guidelines for Common Access to Biological Resources and Information (CABRI), which were published under the title "Laboratory procedures for microorganisms. M/1998/3.00 Appendix 5.07, Cryopreservation of fungi". They recommend that cryopreserved organisms should preferably be stored below -135°C in order to be nearly at or below the limit for ice crystal growth (-139°C) and state that storage temperatures above -70°C are not acceptable. The guidelines comprise a lot of very important general rules for preservation protocols, e.g.: "Prior to storage the cells must be cooled at a rate at which production of intracellular ice-crystals is avoided; during freezing the bulk of the free water crystallizes, leaving a saturated solution of high osmotic value surrounding the still unfrozen cells, which dehydrate and shrink to obtain an osmotic equilibrium; the exposure of the cells to this solution should result in dehydration to such an extent that the freezing temperature is depressed sufficiently; the optimal cooling rate depends on the size of the cell, the thickness of the cell-wall and the cryoprotectant".

A multitude of factors affect the effectiveness of cryopreservation in fungi, for example, species, strain, cell size and form, growth phase and rate, incubation temperature, growth medium composition, pH, osmolarity and aeration, cell water content, lipid content and composition of the cells, density at freezing, composition of the freezing medium, cooling rate, storage temperature and duration of storage, warming rate, and recovery medium [11]. One of the most important conditions is the composition of the medium used to suspend the organisms for freezing. Although some workers [e.g. 149] do not use a cryoprotectant (protective additive, CPA) and yet reached a good survival of deep-frozen fungi, the presence of a suitable CPA usually increases the survival considerably. For example, ATCC and CMI use 10% glycerol or 5% dimethyl sulphoxide (DMSO) in the suspending medium. The discovery that glycerol and DMSO protect eukaryotic cells (including certain microbial cells) against freezing damage marked the beginning of modern cryotechnology.

What is very important is the type, concentration, and toxicity of the cryoprotectant. Some of the earliest concepts of cryoprotection were established by biologists in the 19th and early 20th century who studied freezing, cold hardiness and frost resistance in the environment, most often in plants [158]. For example, in 1890's Hans Molisch examined freezing in plants by direct microscopy using an early version of a cryomicroscope. From the recent translation of his original work [159] it is evident that Molisch was aware of the fact that the composition and concentration of substances in plant cell cytoplasm essentially dictated survival or death after freezing. The importance of sugars as CPA was clearly recognized by Maximow in the early 1900's [160]. Similar observations on the biochemistry of cold hardiness in plants and insects continued to be made over the next 30 years, but it was not until the pioneering work of Polge et al. [161] on freeze storage of fowl semen that deliberate addition of a chemical to protect against freezing damage was made; in this case, glycerol was the additive. Following this work, other small molecular weight solutes with high aqueous solubility were shown to possess cryoprotectant properties, prompting the widespread application of cryopreservation in medicine, biotechnology, and plant and animal breeding. In the early years, these protective chemicals were called 'cryophylatic agents' or 'solute moderators', but, following from the lead set by the Society for Cryobiology in 1965, they were given their current designation of 'cryoprotectants' [158]. The influence of cryoprotectant concentration, time of treatment, temperature and pH of the solution is now intensively studied [162]. Similar studies were carried out also in edible mushrooms *Lentinus edodes* [163] and the genus *Pleurotus* [164].

Cryoprotectants are of two types: penetrating agents such as glycerol and dimethyl sulphoxide (DMSO), which readily pass through the cell membrane and protect intracellularly and extracellularly, and non-penetrating agents such as sucrose, lactose, glucose, mannitol, sorbitol, dextran, polyvinyl pyrrolidone, and hydroxyethyl starch, which exert their protective effect external to the cell membrane. Glycerol and DMSO have proved to be most effective for fungi, although polyethylene glycol, another penetrating agent, can be used also [165]. A considerable number of chemically different compounds has been tested for the role of cryoprotectant. An excellent and very useful overview of the main experimental findings with a wide range of cryoprotectants and organisms was published by Hubálek [42]. This review updates the corresponding chapter in his fundamental book publication on cryopreservation of microorganisms [11]. Another useful review article on cryoprotectants was published by Smith [43].

The main categories are sulphoxides (e.g. dimethyl sulphoxide), alcohols and their derivatives (e.g. methyl alcohol, ethyl alcohol, polyvinyl alcohol, ethylene glycol, propylene glycol, polyethylene glycol, polypropylene glycol, polyethylene oxide, glycerol), polyalcohols (e.g. mannitol, sorbitol), saccharides (e.g. glucose, xylose, saccharose, lactose, maltose, trehalose, raffinose), polysaccharides (e.g. dextran, mannan, dextrin, starch), amides, N-alkylamides and imides (e.g. acetamide, methyl acetamide, dimethyl acetamide, dimethyl formamide, succinimide), heterocyclic compounds (e.g. methyl pyrrolidone, polyvinyl pyrrolidone), amino acids and carbonic acids (e.g. glycine, glutamic acid, glutaric acid, ammonium acetate, EDTA), proteins, peptides and glycoproteins (e.g. serum, albumins, gelatine, peptones, glycoproteins), complex substrates (e.g. yeast extract, malt extract, skimmed milk) and some surfactants (e.g. Tween-80, Triton). Combination of several cryoprotectants turned out to be useful, e.g. the mixture of 10% DMSO with 8% glucose in case of some fastidious fungi [43].

As mentioned above, the cryoprotective additives (CPAs) used in the frozen storage of microorganisms include a variety of simple and more complex chemical compounds, but only a few of them have been used widely and with satisfactory results: these include dimethyl sulphoxide (Me_2SO), glycerol, blood serum or serum albumin, skimmed milk, peptone, yeast extract, saccharose, glucose, methanol, polyvinylpyrrolidone, sorbitol, and malt extract. However, diols (as well as some other CPAs) are toxic for many microbes. The best CPA, or combination of CPAs, and the optimum concentration for a particular cryosensitive microorganism has to be determined empirically [11].

Sakurada et al. [166] examined the cryopreservation of an anaerobic rumen fungus, *Piromyces communis* OTS1, at -84°C using dimethyl sulphoxide, propylene glycol or ethylene glycol as cryoprotectants. Ethylene glycol was the most effective agent (a survival of 80% after 1-year storage), combining high survival and low toxicity, followed by dimethyl sulphoxide and propylene glycol. Cell-free rumen fluid in the cryopreservation medium decreased the toxicity of the cryoprotectant agents and also had a protective action per se.

In many organisms there are naturally present antifreeze proteins (AFPs), which have in some respects similar effect like cryoprotectants. These proteins have the ability to modify the growth of ice, resulting in the stabilization of ice crystals over a defined temperature range and in the inhibition of the recrystallization of ice. AFPs are found in a wide range of organisms, including bacteria, fungi, plants, invertebrates and fish. Moreover, multiple forms of AFPs are synthesized within each organism [167].

The cryopreservation process includes freezing and thawing, the rate of both these processes being important [20, 168, 169]. Many fungi survive rapid and uncontrolled freezing (direct immersion in LN, representing several hundred °C/minute), other less cryoresistant fungi require controlled, slow-rate freezing [18, 145, 170, 171, 172]. A considerable example was demonstrated by Mazur [173, 174], who found that survival of ~~of~~ fungal cells suspended in water could be increased from less than 1% to nearly 100% just by modifying the cooling and warming rates. A large number of cryopreservation protocols has been developed and published [see 11] showing that very different procedures are required by individual fungi. Generally, for the majority of fungi the best procedure is ~~a~~ slow cooling combined with ~~a~~ rapid thawing. Usually a cooling rate of 1°C per minute to -40°C is applied (but not for all fungi; some are damaged by this cooling rate, e.g. *Oomycetes*). The frozen cultures are revived by warming them as fast as possible. Generally the cryovials are thawed in a water bath at 25-30°C [175]. The application of controlled rate cooling with cryoprotectants lessens the risks of cryoinjury [176]. Smith [43] suggests that cryoinjury in fungi is a result of the interaction of several stresses, such as the rupturing of tissue by ice damage or concentration/solution effects. The latter includes pH changes caused by precipitation of buffers, dissolved gases, electrolyte concentration, intracellular crystallisation resulting from loss of the water of hydration from macromolecules, and cell shrinkage [177]. Membrane damage may be a result of solution effects, but could also be caused by ice damage. Indeed, Roquebert and Bury [163], who studied the effects of cooling on *Lentinula edodes*, suggest that ice crystals may exhort a physical destructive pressure on membranes causing death. Freezing damage as described by Mazur [178] is a balance of two effects, dehydration, which he calls solute effect (occurs when cooling is very slow) and intracellular ice formation (when freezing is relatively fast). This latter can be overcome by ultra-fast cooling, but recrystallization is possible if the material is allowed to re-warm and rest at temperatures of about -40°C. Slow cooling at a rate of between 1 to 10 degrees per minute is recommended especially over the eutectic point (freezing point). It is recorded [3, 178] that if slow cooling is used, the rate of thawing has little effect, but if rapid cooling is used, rapid thawing is essential. Unfortunately, the research confirmed that cryoinjury can proceed in the course of the whole process (cold shock, osmotic shock, freezing injury, storage injury, thawing injury, recovery injury).

For fungi sensitive to cryoinjury and for those which exhibit poor viability following preservation, specific methods may need to be developed to ensure optimal cryopreservation. Examples of these are some members of the *Basidiomycota* and *Straminipila* such as *Phytophthora* and *Saprolegnia* spp. Such organisms are often termed preservation recalcitrant and finding a suitable method for the long-term preservation of such fungi can be difficult. For laboratories with limited funds and equipment, the required controlled rate of -1°C per minute can be achieved without the need for a controlled rate cooler using the commercial ´Mr Frosty´ (Nalgene, Rochester, NY) bench top cooler. Also the desirable temperature of liquid nitrogen (-196°C) or its vapour phase (about -130°C), which suppress all chemical activity, is not strictly required as the death rates at about -80°C are very low [35, 179, 180].

Spores are generally more cryoresistant than vegetative forms [136]. This is due to a lower water content and their usually thicker cell wall. For instance, all spore types (urediniospores, teliospores, basidiospores) of rust fungi are much more resistant to freezing and thawing than mycelial culture [181]. Even the different forms of mycelium differ in their sensitivity to freezing. Tertiary hyphae of the basidiomycete *Lyophyllum ulmarium* were lethally injured by slow cooling and freezing in LN, whereas secondary hyphae tolerated this treatment well [182]. Low water content (e.g. in dry spores) generally increases the resistance to freezing due to reduced possibility of intracellular ice formation [178]. Also the cultures from the stationary phase of growth are generally more resistant to cryoinjury than those from the exponential phase. E.g., logarithmic phase of mycelial *Neurospora crassa* has been extremely sensitive to all rates of freezing and thawing studied, while at the stationary phase mycelium showed slight cryotolerance [183]. Young, actively-growing hyphal tips of *Neurospora* have been found to be less resistant to freeze-thaw injury than older mycelial parts or spores [184]. The sensitivity of fungi to cryopreservation can also be influenced by the composition of the growth medium, its pH or osmolarity, nutrient limitation, fluidity of the cell membranes, vacuolization etc. [e.g. 185]. The presence of salts can be damaging due to ice formation around the cells, which was demonstrated on viability of *Neurospora crassa* conidia in water suspensions [186]. The cultivation medium is usually not very convenient for cryopreservation. Inspirational synopsis on tolerance to freezing and thawing published Smith [17]. He pointed out to the capability of many fungi to tolerate very low temperatures, especially in the form of spores. E.g., large numbers of soil-borne fungi survive severe winters. The vegetative hyphae are more sensitive to freezing than the structures produced for propagation or for withstanding adverse conditions such as overwintering, lack of nutrients, or drought.

An overview study [1] on the influence of cryopreservation process on survival of taxonomically very broad spectrum of fungi showed that there was no obvious link between taxonomic grouping and the response of the fungi to freezing and thawing. Nevertheless, the number of characteristics evaluating the success of preservation should be increased, because the complete revival of cryopreserved cultures (evaluated mostly by measuring the colony diameter) is generally still uncertain. The survival rate varies between 60 and 100% [73, 163, 164, 187, 188], which indicates a need to optimize the current cryopreservation methods and to develop new ones.

More recently, further procedures have been developed for preservation-recalcitrant fungi that otherwise could not be preserved; these techniques include *encapsulation and vitrification-based cryopreservation* approaches. *Vitrification* is a relatively new cryopreservation technique that has been according to the present interpretation used since 2000. Nevertheless, by the year 2000, a total of 74 reports were recorded in the ISI databases. Of these, the majority (60%) concerned use in storage of embryos and oocytes; the second place took cell and tissue preservation (25%). This compares to only 8% of recorded activity for plant cells in 2000 in conventional cryopreservation, and indicates that vitrification has led to an important expansion in low temperature technology for applied plant breeding and biotechnology. The method involves rapid freezing of cells that has been treated with very highly concentrated cryoprotectant solutions (usually sugar mixtures) in order to produce an amorphous glass that surrounds the cells. This prevents the onset of "concentration effects" or ice damage [117]. The composition of the cryoprotectant solution is important since the protectants must remain liquid at extremely low temperatures. But many of the solutions that remain liquid and are able to permeate the cell wall are toxic in high levels and can damage the cell. A lot of cryoprotectants are being tested for vitrification solution, e.g., propane diol, ethylene glycol, glycerol, sucrose and other sugars, dimethyl sulphoxide, ethane diol, sorbitol, lactalbumine and many others. Samples are cooled directly in liquid nitrogen which eliminates the need of the controlled rate cooling. On resuscitation from the frozen state, care must be taken to ensure that samples do not ′crack′, which could cause physical damage to the mycelium. More importantly, samples must be immediately washed to remove the toxic vitrification solution. The technique has been applied to a range of fungi with some success [189]. Tan and Stalpers [190, 191] obtained satisfactory results with the fungus *Arthrobotrys superba* CBS 643.80. Cells were suspended in a 2 M mixture of permeating cryoprotectants. Once the cryoprotectant mixture entered the cells, they were exposed to an 8 M mixture of cryoprotectants. After sufficient dehydration, cells were plunged in liquid nitrogen and cell content as well as cryoprotectants were vitrified.

However, vitrification solutions are toxic to most living organisms, so the routine use may not be advantageous. *Immobilisation (encapsulation cryopreservation)* is a technique where cells are encapsulated in calcium alginate beads prior to preservation [192]. Examples of the use of this technology are now well documented for fungi; for example *Serpula lacrymans* [193] and monoxenically produced spores of *Glomeromycota* [194]. The application of encapsulation has two main functions: firstly, it allows cells to be easily manipulated by providing a suitable suspending matrix; secondly, the water content of the cells can be reduced by osmotic treatment or drying, which decreases the prospects of ice damage or concentration effects during the cooling stage of the cryopreservation procedure. Vitrification and immobilisation cryopreservation techniques have great potential for preserving recalcitrant fungi that would otherwise not be stored for long. However, these methods are very time consuming and are still relatively under-researched and therefore not broadly tested.

The concept of bead encapsulation has become highly relevant to agriculture. This procedure enables to temporarily protect the encapsulated microorganisms from the soil environment and microbial competition and to release them gradually for the colonization of plant roots [195].

A very useful tool in morphological research in cryobiology is *cryomicroscopy* [125, 176, 196, 197, 198]. This method allows the direct observation and recording of changes during freezing and thawing. Smith and Thomas [176] observed 195 strains of fungi during freezing and thawing using a cryogenic light microscope. It is important that they did not find any obvious link between taxonomic position and the morphological response to freezing and thawing. The viability of 7 of these strains was examined following freezing and thawing in the presence or absence of the cryoprotectants glycerol and dimethyl sulphoxide. Intracellular ice and hyphal shrinkage were not necessarily lethal events, but in many cases they affected the rate and quality of growth. Both cryoprotectants reduced shrinkage, shifted the cooling rate where intracellular ice formed in many cases, and improved the recovery of strains. These studies enable to optimize cryopreservation protocols.

Cryomicroscopy is also a useful technique for studying interactions between plant and fungal pathogen. A new method of sample preparation for cryo-scanning electron microscopy was used to visualize internal infection of wheat (*Triticum aestivum*) roots by the pathogenic fungus *Rhizoctonia solani* AG-8 [199]. This method retained fungal hyphae and root cells in situ in disintegrating root tissues, thus avoiding the distortions that can be introduced by conventional preparation by chemical fixation, dehydration and embedding. Infected roots frozen in liquid nitrogen were cryo-planed and etched (sublimed) at -80°C for a critical length of time (up to 9 minutes) in the microscope column to reveal plant and fungal structures in three dimensions. Root and fungal structures were well preserved irrespective of the infection severity. Root and hyphal cell walls were clearly seen and hyphal architecture within and between root cells was preserved. This rapid method permits three-dimensional *in situ* visualization of fungal invasion within roots and has broad application for examination of diseases caused by other necrotrophic fungi.

Research has been focussed on improvements in methodology and *assessment of stability* of specimens in culture. As mentioned above, some fungi are preservation recalcitrant and difficult to store for long-term. Recalcitrant fungi include generally unculturable obligate fungi such as microcyclic rusts, straminiples such as *Halophytophtora*, *Saprolegnia*, and *Aphanomyces* spp., some *Basidiomycota*, and the *Glomeromycota*. Research aims to develop improved cryopreservation techniques for such organisms, which are frequently of economic significance. The techniques employed include vitrification cryopreservation [200], immobilisation [193], and cryopreservation with the host substrate [201]. An alternative approach to cryopreservation of air-dried conidia, notably of the grapevine powdery mildew *Uncinula necator* used Stummer et al. [21].

The maintenance of phenotypic and genomic integrity is crucial, but this is difficult for some fungi. The fungi are generally rather variable due to dimorphism, the presence of extranuclear genetic elements (e.g., transposons), mutations, etc. Variation in culture morphology may also be exhibited following preservation by non-optimal preservation methodology. Despite the unstable nature of some fungi, the aim of preservation is to ensure long-term survival of an organism without a change in its physiological and genomic integrity [189]. Viability cannot be the only criterion for the assessment of the cryopreservation success. Many authors have found differences in post-preservation stability following the maintenance of cultures in the culture collections. RFPL fingerprints revealed polymorphism and chromosomal abnormalities associated with culture degeneration of *Agaricus bisporus* [202]. Kuhls et al. [203] pointed out that presumably identical strains of *Trichoderma* obtained from different collections had different PCR fingerprints. Also Broughton et al. [26] assessed genomic integrity in five strains of the economically important fungus *Trichoderma*. They applied two PCR fingerprinting techniques and detected amongst replicates three post-preservation genetic polymorphisms, indicating that even standardized cryopreservation methodologies can sometimes induce genomic change. PCR fingerprints of one strain of *Fusarium oxysporum* f.sp. ciceris maintained for 12 years differed from the typical fingerprints of other strains of the same species, indicating that the strain deteriorated during the maintenance. Ryan et al. [204] found that sector formation was associated with changes in physiology in a strain of *Metarhizium anisopliae*. Changes in physiological features following preservation were also found in *Penicillium expansum*. The production of patulin and citrinin, the secondary metabolites of the fungus, was altered [205]. Even worse, PCR fingerprinting of replicates of an isolate of *M. anisopliae* subjected to non-optimized preservation techniques revealed evidence of polymorphisms [20]. It is especially important to ensure the stability of the strains used for biotechnological applications [200, 206]. The effects of cryopreservation protocols on pathogenicity have been assessed for entomopthoralean fungi [70, 90, 207] and *Fusarium moniliforme* [208], while the influence of different protocols on the fruit body production of cryopreserved test cultures of *Flammulina velutipes* [209], *Agaricus bisporus* [210] or *Agaricus brunnescens* [211] was also reported. The effect of cryopreservation and freeze-drying on the instability of the strains of filamentous fungi was compared by Dupont et al. [212]. They reported an exceptional finding for *Trichoderma viride*, where freeze-drying provided better results than refrigeration in LN.

Unlike freeze-drying, LN refrigeration does not stimulate any mutagenicity [213]. The two phenotypes of *Histoplasma capsulatum*, A (albino) and B (brown) had not changed after LN storage, whereas routine subculturing led to transitions from B to A type [214]. The colour of the urediniospores of *Puccinia graminis* f. tritici was preserved after storage in LN, contrary to other maintenance methods [2], and a u.v.-induced mutant of *Ustilago violacea* remained stable after LN refrigeration [215]. No morphological or physiological changes have been observed in 7354 fungal strains stored in LN at the International Mycological Institute, England [63]. MacDonald [216] reported a safe keeping of conidia of mutant strains of *Penicillium chrysosporium*. Also production abilities, drug sensitivity and other characteristics of the strains used in biotechnology do not change after LN refrigeration, e.g., penicillin and other metabolite production [13, 24, 184, 185, 216, 217].

Reports on cryopreservation of *specific genera or groups of fungi* are often published. Problematic organisms include obligate symbionts or parasites [27], some members of the *Basidiomycota* [27, 218] and some other groups of fungi. For the first group, Declerck and Van Copennolle [194] reported the successful cryopreservation of *Rhizophagus* sp. (formerly *Glomus intraradices*) MUCL 41835, an arbuscular mycorrhizal (AM) fungus, while Ryan and Ellison [201] succeeded in the ultralow preservation of *Puccinia spegazzinii*, a microcyclic rust fungus.

In the central bank of the microorganisms section in the Ministry of Agriculture, Forestry and Fisheries (MAFF) Genebank in Tsukuba a survival of fungi and *Oomycetes* was evaluated. 6,578 strains out of 6,681 tested strains (98.5%) and 264 (63.8%) of 414, respectively, survived one-year preservation in the vapor phase of LN. The details on survival of the preserved fungal strains were as follows: 99.0% (1,107 strains/1,118 tested strains) survival in *Ascomycota*, 96.1% (1,552/1,615) in *Basidiomycota*, 98.6% (73/74) in *Zygomycota*, and 99.3% (3,846/3,874) in anamorphic fungi. Thus, cryopreservation is excellent for preservation of most fungal strains, but is not suitable for preservation of many oomycetous strains [33]. Cryopreservation in LN was successfully used even for marine thraustochytrids [219]; Yarlet et al. [220] cryopreserved the anaerobic rumen fungus *Neocallimastix patriciarum*.

What is very important is the preservation of pathogenicity of *phytopathogenic* (especially rust) fungi [2, 181, 221, 222, 223, 224, 225, 226] and *human or animal pathogenic* fungi [227, 228] including *entomopathogenic fungi*, e.g., in the form of protoplasts of *Entomophthora* [229].

Numerous pathogenic fungi must be cryopreserved with their host or alternative growth substrate. This approach has been applied in various forms for many years. For example, plant seeds have been used to support members of the *Straminipila* when cryopreserved. For obligate pathogens or mutualists, cryopreservation has been attempted where the fungus is preserved with its host substrate. This method has been used for many recalcitrant organisms, such as the microcyclic rust *Puccinia spegazzini* where the teliospores were preserved on the stem or petiole tissue [201]. Wood et al. [230] used a combinational approach of cryopreservation with a substrate and encapsulation cryopreservation. Seeds of the common spotted orchid (*Dactylorhiza fuchsii*) and green-winged orchid (*Anacamptis morio*) were encapsulated in alginate beads with hyphae of the basidiomycete *Ceratobasidium cornigerum* with no post-cryopreservation adverse effects. The combination of cryopreservation with an appropriate substrate/encapsulation cryopreservation approach may also be suitable for the cryopreservation of other mutualistic associations such as endophytes and lichens. E.g., cryopreservation was applied in lichen-forming ascomycetes and their green algal photobionts [231]. They completely die off within approximately 3 years of storage at room temperature. Macroscopically this is recognizable as a colour change, the green shades of the chlorophylls being lost. In fluorescent light microscopy preparations an increase in fungal autofluorescence and a significant decrease in chlorophyll autofluorescence in the *Trebouxia* cells was observed. In transmission electron microscopy preparations of *Xanthoria parietina* and its green algal photobiont, *Trebouxia arboricola*, the fungal membrane systems were found to be largely broken down whereas the shrivelled algal protoplast failed to rehydrate after storage at room temperature. When stored in the desiccated state at -20°C, both partners of the symbiosis stayed fully viable for up to 13 years, their colouration and chlorophyll fluorescence being unchanged. In the desiccated state deep-frozen samples can be repeatedly brought to room temperature and back to -20°C without any loss of viability.

These alternative approaches to cryopreservation have great potential for a large number of unculturable fungi that are otherwise not maintained in a living state. Until now, such fungi have only been stored as air-dried herbarium specimens [27].

Phytopathogenic fungi are difficult to maintain in pure culture without loss of pathogenicity, virulence, and the ability to sporulate. Dahmen et al. [18] studied survival, pathogenicity, and sporulation of 31 phytopathogenic fungi (including species of *Myxomycetes*, *Oomycetes*, *Ascomycetes*, *Basidiomycetes*, and *Deuteromycetes*) after the storage in liquid nitrogen for up to 9 years. The protective solutions used and rates of freezing and thawing were crucial for successful long-term preservation. Some fungi survived in water as well as in a protective solution. Skim milk-glycerol (8.5% + 10%) and DMSO (15%) were good cryoprotectants for all propagules tested except urediospores of rust fungi, which required dry storage. All but one (*Erysiphe graminis*) of the 31 fungal species studied could be preserved for up to 9 years. Ochi and Nakagawa [232] developed a simple method for long-term preservation of the soybean red crown rot fungus, *Calonectria ilicicola*, using barley grains. The fungus preserved on barley grains at -80°C remained viable without any change in mycelial growth and virulence for at least 3 years. Different storage methods were tested to examine their usefulness in viably maintaining field isolates of *Cochliobolus sativus*. Fungal isolates were stored in sand, in silica gel and on potato dextrose agar (PDA) medium at 4°C or frozen as conidia at -20°C. The cultures were assessed after 12 and 24 months of storage. Cultures of all isolates were viable after two years of storage in sand and in silica gel at 4°C or dry at -20°C, whereas isolates stored on PDA were not viable and heavy contamination with bacteria was frequently observed. After 2-year storage, isolates were able to infect barley leaves, but sporulation was reduced dramatically, depending on the isolate and the storage method employed. AFLP analysis revealed several genetic changes between stored and non-stored isolates. A total of 444 scorable DNA bands were obtained; 324 of these (71%) were polymorphic. However, the proportion of DNA alterations varied according to the storage method [25]. Viability and infectivity of dry harvested spores of *Puccinia abrupta* var. partheniicola were successfully preserved by cooling to and storage at -190°C. Spores remained viable for 32 days following thawing and were able to cause normal infections. The fungus as infected tissue supporting spores was less readily preserved. When the cryoprotectants glycerol, dimethyl sulphoxide, trehalose and polyvinyl pyrrolidone were added prior to freezing, both viability and infectivity were reduced [233]. Cryopreservation of the biotrophic rust fungus *Puccinia spegazzinii* (*Basidiomycotina*, *Uredinales*), which has been identified as a potential classical biological control agent for the invasive weed *Mikania micraniha* (*Asteraceae*), was only partially successful [201]. Infection of the host plant by basidiospores produced from previously cryopreserved teliospores, embedded in leaf petioles, was not achieved.

Legard and Chandler [234] evaluated the survival of several strawberry fungal pathogens in a -95°C ultra-low temperature freezers. All cultures of *Colleotrichum fragariae*, *C. gloelosporioides*, *C. acutatum* and *Phomopsis obscurans* survived long-term storage in excellent condition. They evaluated also the effect of rapid, uncontrolled freezing in the -95°C freezer and of thawing at room temperature. This procedure was repeated successfully four more times without noticeable changes.

The stability of resistance to fungicides is one of the features of interest in preservation of plant pathogenic fungi. Zhu et al. [235] investigated twelve storage approaches to determine their effects on instability of propiconazole resistance for three demethylation inhibitor (DMI) fungicide-resistant and two DMI-sensitive isolates of *Monilinia fructicola*. They included mycelium in PDA slants under mineral oil, in PDA plugs under 10% glycerol, mycelium on dried filter paper and conidia on silica gel, each stored for 36 weeks at 4, -20, and -80°C. None of the storage approaches prevented the rapid decline of EC50 values for propiconazole in the three resistant isolates, and no significant differences were found among storage approaches ($P = 0.787$) or between storage approaches and consecutive transfers ($P = 0.053$). Most of the decline in resistance occurred during the first 4 weeks of storage.

Rhizoctonia solani is another important plant pathogen for a number of crops and maintaining an extensive collection of reference isolates is important in understanding relationships of this pathogen with multiple hosts. Webb et al. [236] examined the potential of cryopreservation in vapour phase of LN for storage of a *R. solani* culture collection containing 106 isolates (primarily from sugar beet). After 60 days, 5 years and 10 years in storage, all isolates were tested for viability by calculating the percentage of barley grains from which *R. solani* mycelia grew. Five years after the initial storage, all isolates except one had no change in viability. After 10 years in storage, 67 of 106 isolates had no significant decrease in viability, 39 of 106 isolates had a significant decrease in viability but only 9 isolates had less than 10% growth, with 4 having no growth. A subset of isolates stored for 10 years were tested for pathogenicity on a susceptible (FC901) and resistant (FC703) sugar beet germplasm. All isolates tested maintained approximately the same level of virulence that they had prior to storage on both germplasms.

Conidia of *Uncinula necator*, the causal agent of grapevine powdery mildew, can be stored at -70°C for up to 12 months following air drying without cryoprotectants. After recovery they are able to infect surface-sterilised, detached grapevine leaves. Conidia also survived five freeze/thaw cycles. Southern blot analysis revealed that DNA extracted from colonies resulting from cryopreserved and fresh conidia produced the same DNA fingerprints [21]. *Podosphaera fusca* is the main causal agent of cucurbit powdery mildew in Spain and one of the most important limiting factors for cucurbit production worldwide. As an obligate biotrophic parasite, this fungus has been traditionally cultured and conserved by periodical transfers of conidia to fresh plant material. Perez-Garcia et al. [237] described a simple protocol for preservation of *P. fusca* isolates in absence of living tissue based on the dry spore, slow-freezing technique, and demonstrated that storage of silica gel desiccated conidia at -80°C is an efficient method for the long-term preservation of the pathogen. In the course of the effort to find preservation techniques for three *Rosellinia* isolates used in plant pathogenic research, ten Hoopen et al. [238] found out that development of a generalised procedure for storage of the three *Rosellinia* species (*R. bunodes, R. necatrix* and *R. pepo*) was not possible and successful storage protocols had to be developed for individual isolates.

Some obligate parasites can be propagated only in planta, which necessitates development of a method for preserving cultures for longer periods of time for routine biological studies and shipment to other laboratories. Widmer [239] studied preservation of the fungus *Synchytrium solstitiale*, a candidate for use as a biocontrol agent against *Centaurea solstitialis*. The best results were obtained when infected tissue was stored at -2ºC in 0.5 M sucrose where active zoospores were released after 12 weeks of storage.

A very specific group are *entomopathogenic fungi* (*Entomophtorales*). Viability of isolates from three species of this group (*Entomophaga aulicae*, *Entomophaga maimaiga* and *Zoophthora radicans*) stored at -80°C after a simple freezing protocol was comparable to that of isolates stored at -196°C [207]. Three isolates of *Entomophaga maimaiga*, a fungal pathogen of gypsy moth (*Lymantria dispar*), were stored immersed in LN (-196°C) as naturally occurring protoplasts for approximately 7 years. All isolates survived although with diminished ability to produce conidia and azygospores. Days from infection to host death increased for all isolates after the storage period. Soon after isolation, two of the isolates differed in levels of azygospores produced at 25°C, while after storage, differences among isolates persisted but had changed; at 25°C, all isolates differed from each other in levels of conidial production, and one isolate no longer produced azygospores. Days from infection until host death for one isolate were significantly longer in comparison with other isolates both before and after storage, and this latter isolate also grew slowly *in vitro* as well as *in vivo* [90]. Shimizu and Hayata [240] used for cryopreservation protoplasts of an entomogenous fungus, *Paecilomyces fumosoroseus*. For preservation of mycelia of the entomopathogenic fungus *Zoophthora radicans* isolate ARSEF 1590 (*Zygomycetes*, *Entomophthorales*), in LN, three cooling rates (slow, intermediate, and ultrarapid) and four cryoprotectants (glycerol, dimethyl sulphoxide (DMSO), 1,2-propane diol, and methanol) were tested. The best combination was the slow cooling rate and 5% 1,2-propane diol with 89% survival after thawing [241]. Twenty four strains of the entomopathogenic fungi (*Hyphomycetes*) *Beauveria bassiana*, *Metarrhizium anisopliae*, *Nomuraea rileyi*, *Paecilomyces farinosus*, *P. fumosoroseus* and *P. lilacinus* were preserved by lyophilization and in LN and their conidial viability was assessed. Germination rates of 16- to 84-month-old cultures stored in LN decreased, on average, by less than 13.3%. For 29- to 49-month-old cultures preserved by lyophilization, the viability loss ranged, on average, from 28.6 to 94.5% [242]. *Beauveria bassiana* is a worldwide distributed entomopathogenic fungus, which is used nowadays as an alternative to chemical pesticides in the control of several agriculture pests. Oliveira et al. [243] evaluated the effect of 1-year preservation by lyophilization, glycerol-freeze at -20°C and subculturing on growth, production and viability of spores and macro- and micro-morphology of three isolates of this fungus. For long-term storage, glycerol-freeze proved to be the most capable method of preservation. Preservation of entomopathogenic fungi was described by Humber [244].

Cryopreservation can be used as well for the storage of artificially prepared fungal protoplasts [245, 246]. An electroporation method, utilizing cryopreserved protoplasts, has been developed for the steroid 11-hydroxylating fungus *Curvularia lunata* strain IM 2901 [247]. Protoplasts released from the apical parts of 24- and 48-h-old mycelia were suspended in cryopreservation buffer and stored at -75°C for several weeks. The thawed and freshly prepared (control) protoplasts were electroporated with pAN 7-1 plasmid carrying the *Escherichia coli* hygromycin B resistance gene (hph) under the control of *Aspergillus nidulans* sequences. The electroporation efficiency of the control protoplasts with plasmid pAN 7-1 was 7.5 and 12.0 transformants per mg DNA (protoplasts liberated from 24- and 48-h-old mycelia, respectively). Protoplasts released from the younger mycelium were more stable according to their reversion ability to mycelial form and transformation efficiency. After 16 weeks of cryopreservation the yield of electroporation was 61.3 % of the control value. All isolated electrotransformants proved to be stable for a period of over 4 months even without selective pressure.

Nematophagous fungi can be useful in the animal production. They produce special structures along the hyphae, named traps, which are responsible for capturing and killing the nematode pre-parasitic stages. The biological control of helminth parasites of bovines by these fungi is an alternative to the use of drugs with the principal objective of reducing the source of infection available on pastureland. The maintenance of predatory activity of the fungal isolates is one of the basic prerequisites to ensure the success of this form of control [248, 249]. In their study behaviour of isolates I31 of *Arthrobotrys robusta* and NF34a of *Monacrosporium thaumasium* was investigated following three storage methods: stored at 4°C, cryopreserved with or without cryoprotectants or preserved in silica gel. All samples were subsequently passed through the gastrointestinal tract of calves. The latter involved the administration of 20 g of mycelia to the animals. This quantity was sufficient to recover fungal material from the faeces. The peak reduction in the number of infective larvae in the faeces occurred 24 h after administration of the samples ($P < 0.05$). The storage at 4°C was the treatment that produced the greatest reduction in larvae for NF34a (81.3%) and I31 (65.1%) isolates. Nf34a isolate was responsible for the highest percentage reduction of larval helminth populations ($P < 0.05$). Cryopreservation appears to be an efficient method of preserving isolates, although diminished predatory capacity compared to storage at 4°C was seen only for isolate NF34a (73.2%). Cryopreservation did not interfere in predatory activity of I31 isolate ($P < 0.05$). Maintenance of isolates in silica gel showed the lowest reduction throughout the experiment ($P < 0.05$).

Dermatophytic fungi can present another problem for storage, as their cultures often become pleomorphic, with various levels of sporulation or mycelial growth. The degree of sporulation and the quality of the inoculum appeared to be critical factors; when the inocula were "adequate" in size, even some poorly sporulating species such as *Trichophyton violaceum*, *Trichophyton schoenleinii*, and *Microsporum ferrugineum* survived the storage well. Commercially prepared cryogenic tubes (Microbank vials, Prolab Diagnostics, Austin, Texas) were found to be particularly appropriate for long-term preservation of a wide range of fungi, including dermatophyte isolates, at -70°C [133, 250]. Since 1994, 6,198 yeasts and 391 molds belonging to 25 and 37 species, respectively, were stored in these vials at ≥-130°C in LN and at -70°C in a freezer. All of the isolates, with the exception of 45 yeasts and 15 dermatophytes, were recovered from both storage temperatures. Good reproducibility of sensitivity was demonstrated for amphotericin B, fluconazole, and voriconazole MICs determined for random isolates. Espinel-Ingroff et al. [251] used the Microbank system, originally developed for storage of bacterial cells, for the preservation of pure cultures of 391 molds including dermatophytes *Histoplasma capsulatum, Blastomyces dermatitidis,* and *Alternaria* spp. The system consists of sterile vials that contain 25 porous, colored beads and a cryopreservative fluid. Two temperatures, ≥-130°C (LN vapor) and -70°C (freezer), were evaluated. Nevertheless, the Microbank system appeared to be inconvenient for preservation of dermatophytes. On the contrary, Desmukh [96] successfully used simple methods (maintenance on agar slopes in a refrigerator at +8°C and in sterile distilled water, under liquid paraffin, dried on silica gel, in sterile soil at room temperature) for preservation of a total of 239 fungi belonging to 9 genera and 23 species of keratinophilic fungi and related dermatophytes. According to Hasegawa [252], the storage in LN preserves the sporulation ability of fungi and prevents the pleomorphic alteration of mycelial fungi like dermatophytes.

Baker and Jeffries [250] stressed that dermatophytic fungi can present problems for storage, as the cultures often become pleomorphic, with various levels of sporulation or mycelial growth. Although McGinnis et al. [85] confirmed the possibility of using sterile distilled water to store hyphal and spore suspensions of at least 25 species of dermatophytes, they decided on cryopreservation and supported the findings of Chandler [133] that commercially prepared cryogenic tubes (Microbank vials) stored at -70°C are particularly appropriate for long-term storage of a wide range of fungal species, including dermatophyte isolates. Pasarell and McGinnis [128] determined viability of 1,447 clinical and environmental isolates of molds, yeasts, aerobic actinomycetes, and algae belonging to 164 genera (382 taxa) maintained on potato dextrose agar at -70°C for periods ranging from 6 months to 13 years. 33 isolates did not grow after repeated attempts at subculturing. Neither time in storage nor taxonomic classification was associated with a lack of viability.

A lot of fungal diseases afflict *animals*. For example, *Batrachochytrium dendrobatidis* is a major pathogen of frogs worldwide. It has been associated with catastrophic declines of frog populations including those in pristine habitats in Queensland, Australia. To facilitate genetic and disease studies of this fungus and related species, it is essential to have a reliable long-term storage method to maintain the genetic integrity of isolates. Boyle et al. [228] have adapted well-established techniques used for the long-term storage of tissue-culture cell lines to the preservation of *B. dendrobatidis* and other chytridiomycetes. This simple method has allowed to recover these fungi from storage at -80°C and in LN over an extended period. With this technique it is now possible to preserve saprobic and parasitic isolates from a variety of environmental and disease situations for comparative genetic and biological studies. Delalibera et al. [253] investigated the efficacy of several cryoprotectants at different exposure times, cooling rates, and warming rates for protecting hyphal bodies of *Neozygites tanajoae*, a new fungal pathogen used as a classical biological control agent against the cassava green mite, *Mononychellus tanajoa* during cryopreservation at ultra-low temperatures of -80°C or -196°C. They concluded that the optimal cryopreservation protocol differs significantly from the protocols routinely used for relative fungi. Long-term storage of *Ascosphaera aggregata* and *A. apis* causing two important fungal bee diseases tested Jensen et al. [254]. Cryopreservation of hyphae was successful only in case of *A. apis*, while freeze-drying was successful in case of spores (but not hyphae) of both pathogens.

Methods for the preservation of chytridiomycetes in culture collections reviewed Gleason et al. [255]. Chytridiomycetes can be preserved with varying degrees of success using a number of different protocols including cryopreservation. The survival of these fungi is sensitive to environmental factors such as lack of moisture, high temperatures, high osmotic potential, and availability of oxygen, all of which must be considered in designing preservation methods. The age of the culture at the initiation of preservation appears to be a particularly important determinant of viability. High rates of survival have been reported after cryopreservation of aerobic and anaerobic chytrids in 10% glycerol or dimethyl sulphoxide as cryoprotectants.

Cryopreservation of fungi important for *food industry* is a topical subject of research. *Geotrichum candidum*, an extremely common fungus with a world-wide distribution and the causative agent of geotrichosis, is widely used as a ripening agent in some specific cheese making processes. In order to optimize the cryopreservation of this microorganism, freezing experiments were carried out using a Peltier cooler-heater incubator, which permits to control the temperature downshift from +20 to -10°C in time period ranging from 20 to 40 minutes depending on the experiments. Concomitantly, study of the effect of an industrial ice nucleator protein derived from *Pseudomonas syringae* (SNOMAX (R)) on the dynamic of freezing of *G. candidum* was carried out [256, 257]. Their results showed that the addition of this protein to the microbiological suspension had different complementary effects: (i) synchronization of the different samples nucleation, leading to an homogeneous and earlier freezing, (ii) an increase of the freezing point temperature from -8.6 to -2.6°C and (iii) a significant decrease of the lethality of *G. candidum* cells subjected to a freezing thawing cycles challenge.

Basidiomycetes present specific problems with regard to their preservation, because most of them do not form (or at least not in a sufficient amount of) asexual spores, their dominating form being water-rich mycelium, which is rather sensitive to environmental conditions, and also mycorrhizal relationships play often a certain role. Therefore, the most widely used method of culture maintenance in lyophilized state is rather limited in basidiomycetes, because many of them (especially non-sporulating and mycorrhizal ones) do not survive the process [258, 259]. Most attempts to revitalize dehydrated hyphae of fungi have failed, except for some successes [116, 260, 261]; nevertheless, the real absence of spores must be always carefully checked. Several results show that freeze-drying of fungi can be successful in some cases [73, 82, 118, 262] and the growing interest in this technique can be seen. On the other hand, the method of cryogenic culture maintenance seems to be mostly successful, in particular with non-sporulating cultures [1, 12, 14, 151, 263]. In many collections, e.g. IFO (Institute for Fermentation Osaka), non-sporulating cultures of basidiomycetes are stored by cryopreservation at -80°C in electric freezers [264]. Recent studies of cryopreservation of basidiomycetes are not very frequent and most of them deal with only one or a limited number of species [e.g., 20, 82, 262, 265, 266, 267, 268, 269]. Papers comprising a larger number of species (but never in hundreds) are even more rare [e.g. 1, 14, 19, 64, 258, 270].

The well-known first effective protocol for cryopreservation of basidiomycete cultures is Hwang's protocol [144]. This method was performed as follows: agar-plugs with mycelium were cut randomly from the culture plate with cork-borer and frozen with an appropriate cryoprotectant in the cryovial in LN. Afterwards, this method was modified and improved by some laboratories. Ito and Nakagiri [132] examined viability of 939 strains of basidiomycete cultures in IFO (Institute for Fermentation, Osaka) and NBRC (National Institute for Technology and Evaluation (NITE), Biological Resource Center) , which had been stored in -80°C freezers for 1, 5, and 15 years. The results showed that 830 strains were successfully preserved by freezing agar plugs with mycelium submerged in 10% glycerol as cryoprotectant and storing in -80°C electric freezers for 15 years. In spite of such efforts, many cryosensitive basidiomycete cultures exist, most of which were ectomycorrhizal. Such strains often have been maintained by subculturing or mineral oil storage in culture collections including NBRC. The same classic cryopreservation protocol with agar plugs was used by Chvostová et al. [263] for testing the viability of 252 wood-inhabiting basidiomycete strains of after storage under LN, using 10 % glycerol as a cryoprotectant. 164 strains survived the process: 103 *Aphyllophorales* (out of 138) and 60 *Agaricales* (out of 113). The results indicate that the process of cryopreservation should be modified to achieve a higher survival. Fourteen strains of white-rot basidiomycetes belonging to eight species of two genera (*Inonotus* and *Pholiota*) were tested for their ability to maintain the production of laccase, peroxidase and manganese-dependent peroxidase (enzymes involved in lignin biodegradation) after a short-time preservation in liquid nitrogen with different cryoprotectives (glycerol, dimethyl sulphoxide). No negative effect of cryopreservation or the used cryoprotective on production of the ligninolytic enzymes was found in the fungi tested [13]. Lara-Herrera et al. [268] evaluated viability of 6 mushroom strains of the *Pleurotus* genus (2 from *P. djamor* var. *djamor*, 1 from *P. ostreatus* var. *ostreatus*, 2 from *P. ostreatus* var. *columbinus* and 1 from *P. pulmonarius*) after LN cryopreservation (-196°C). The contact time for the mycelia of these strains with the cryoprotectant (glycerol) was studied 1, 2 and 3 hours before freezing. They also tested the effect of different times (5, 10 and 15 minutes) and temperatures (30, 45 and 60°C) of the thawing system for mycelial recovery. A change in thawing and contact times with the cryoprotectant did not affect the results significantly.

Voyron et al. [22] tested six protocols of cryopreservation at -80°C, and 12 protocols of lyophilization on 15 isolates of white-rot fungi belonging to 10 species. The tested protocols were mainly characterized by the use of different growth media, protectants, time and number of perfusion with protectants and finally by the typology and origin of the samples to be cryopreserved (mycelium/agar plug, whole colony) or to lyophilize (mycelium/agar plug, mycelium fragment, whole colony). Cryopreservation and lyophilization outcomes were checked, at morphological (macro- and microscopic features), physiological (growth rate and laccase, Mn-independent and Mn-dependent peroxidases activities) and genetic level (AFLP analysis). Vitality of all fungi was successfully preserved by all cryopreservation protocols at -80°C, and by two lyophilisation methods. Their results showed that cryopreservation at -80°C did not produce morphological changes in any isolate, while two isolates were affected by lyophilization. None of the physiological features were lost, even though growth rate and enzyme activities were somehow influenced by all preservation methods. AFLP analysis showed that only two isolates that varied in their morphology after lyophilization produced a different DNA fingerprint pattern in comparison with that obtained before lyophilization.

Karaduman et al. [271] compared twelve different methods for storage of a newly isolated mushroom strain, *Schizophyllum commune* Fr. The fungal strain was stored in different media (sterile distilled water, 15% glycerol or media including trehalose) in different forms (mycelium-colonized wheat seeds, mycelium-colonized agar plugs) and at different temperatures (+4, +20, -20°C).

The fungal culture was checked periodically for its viability and biological activity for up to one year. The mycelial recovery, ratio, mycelial growth rate, mycelial biomass weight, and enzyme activity of the recovered cultures were used for comparison of the different methods. After 12 months' storage, the vitality of the strain was successfully preserved and the mycelial growth rates and biomass weight were similar for the different methods of storage. The enzymatic activity showed variability for the different storage methods. Our results show that agar plugs stored in sterile water at +4°C and in 15% (v/v) glycerol at +20°C are preferable methods for the preservation of *S. commune*. Additionally, the methods of mycelium-colonized wheat seeds in sterile water at +4°C and agar plugs in sterile water at +20°C are acceptable alternatives. Colauto et al. [272] evaluated cryopreservation of *Agaricus blazei*, a basidiomycete of medicinal and gastronomic importance, at -70°C using different cryoprotectants and freezing protocols. Malt extract agar disks containing grown mycelia were transferred to cryotubes containing different cryoprotective solutions. The results demonstrated that the cryoprotective agent was the most important variable for maintaining mycelial viability after cryopreservation. A long-term cryopreservation (4 years) was effective when saccharose or glucose was used as cryoprotectant regardless of the freezing protocol, dimethyl sulphoxide was effective when a slow freezing protocol was used, and glycerol, polyethylene glycol, and malt extract were ineffective as cryoprotectants regardless of the freezing protocol. Earlier Colauto et al. [273] tested rice grains as a carrier of mycelium for cryopreservation of *A. blasei* at -80°C.

A new method of cryopreservation using perlite as a carrier of fungal mycelium (perlite protocol, PP) was developed by Homolka et al. [61] and then successfully verified on a large number (442) of basidiomycete strains [14]. Perlite is a unique aluminosilicate volcanic mineral holding and retaining substantial amounts of water, which can be released as needed. The viability and some other characteristics – growth, macro- and micromorphology and laccase production – of 442 strains were tested after 48-hour and then after 3-year storage in LN using a PP. All (100%) of them survived successfully the disposal in LN without noticeable growth and morphological changes. Also laccase production was unchanged. The viability and laccase production of a part (250) of these strains were compared with those of the strains subjected to an original agar plug protocol (OP). Using OP, 144 strains (57.6%) out of 250 survived a 3-year storage in LN. The results indicate that the cryopreservation protocol used significantly influences survival of the strains. Markedly better results were achieved using the PP [14]. In their follow-up research Homolka et al. [24] studied the morphological features, mycelial extension rate, enzymatic activities and possible genetic changes in 30 selected strains of basidiomycetes after 10-year cryopreservation on perlite in LN. They also conducted comparisons with the same strains preserved by serial transfers on nutrient media at 4°C. Productions of ligninolytic enzymes and hydrogen peroxide were studied by quantitative spectrophotometric methods, whereas semiquantitative API ZYM testing was used to compare the levels of a wide range of hydrolytic enzymes. The results showed that cryopreservation in LN did not cause morphological changes in any isolate. The vitality of all fungi was successfully preserved and no physiological features were lost, even though the extension rate and enzyme activity were slightly affected. Moreover, the sequence analysis of eight strains did not detect any changes in their genetic features after cryopreservation.

To find out if potential difficulties caused by interrupted supply of LN or electric power could be overcome by the storage in perlite and also if the used cryopreservation protocol significantly influences the survival of the cultures, viability of 50 mycelial basidiomycete cultures on perlite in cryovials was tested after three successive cycles of freezing, storage in LN, and thawing at 37°C or at room temperature (PP). A great majority of them (47 cultures, 94%) survived this treatment successfully without noticeable growth and morphological changes and also laccase production was unchanged. These characteristics were compared with those of the cultures kept in the form of agar miniplugs in open polypropylene straws placed in cryovials and subjected to the same treatment (straw protocol, SP). Using SP, only 6 cultures (12%) out of 50 survived repeated freezing and thawing. The possibility of reusing the stored cryovials for successive inoculations and direct distribution was tested with success [274]. As mentioned above, perlite can be used for the subcultivation of basidiomycetes [71]. Cultures of 33 basidiomycete strains out of 35 tested were viable with unchanged characteristics after four years of maintenance on perlite in cryovials at 4°C. These cultures can be a good substitute for agar cultures in long-term maintenance of fungi significantly reducing the intervals between transfers.

Sato et al. [275] applied the perlite protocol to the cryosensitive strains in NBRC collection, especially ectomycorrhizal basidiomycete cultures. In this investigation, they selected 7 species (12 strains) of cryosensitive strains, which were not tested before by Homolka et al. [14]: *Tricholoma matsutake* NBRC 6915, NBRC 30773 and *Lactarius hatsudake* NBRC 32777, NBRC 33154 were selected as representatives of unstable strains in long-term cryopreservation by the ordinary agar disc freezing method. The other 8 strains were selected as representatives of those unsuitable for cryopreservation because they showed very low viability after the freeze-thawing. They investigated viability, time to recover, and basic morphological study after freezing and storing at -80°C for 6 months.

The viability of the fungal strains was 60% in *Phallus hadriani* and 100% in remaining 11 strains, indicating the efficacy of PP method for cryopreservation of cryosensitive basidiomycetes. The PP method was modified by limiting the cryoprotectant exposure time to 48 hours, and increasing the glycerol concentration from 5% to 12%. The viability of *P. hadriani* strain increased from 60% to 100% with the modified PP after storage at -80°C for 6 months.

The perlite protocol was tested not only on basidiomycetes, but also on other taxonomic groups of fungi. The viability, growth and morphology of 48 strains of *Ascomycota* (including 17 yeasts) and 20 strains of *Zygomycota* were determined after a 2-day and then after 1-year storage in LN using cryopreservation method with perlite as a particulate solid carrier [62]. In case of *Ascomycota*, 45 strains (94 %) out of 48 survived both 2-day and 1-year storage in liquid nitrogen, respectively. In case of *Zygomycota*, all 20 strains survived both storage treatments. In addition, 3 strains of *Basidiomycota* counted among yeasts were tested and all survived the 1 year storage. In all surviving cultures no negative effects of cryopreservation by this method have been observed after 1-year of storage in LN. The results indicate that the PP can be successfully used for cryopreservation of taxonomically different groups of fungi and also for fungi which failed to survive other routinely used preservation procedures.

A specific approach is required by *mycorrhizal fungi*. Their preservation is generally regarded as rather complicated and uneasy. The reason is that arbuscular mycorrhizae (AM) are symbiotic associations, formed between plants and soil fungi that play an essential role in plant growth, plant protection, and soil quality. AM fungi are ubiquitous in the soil and can form symbiosis with most terrestrial plants including major crops, cereals, vegetables, and horticultural plants. Due to their obligate symbiotic status, AM fungi need to associate with plant for growth and proliferation. These organisms do not produce conidia [276] and the sexual structures are rarely formed on synthetic media. Their most common form *in vitro* is mycelium that generally grows slowly in axenic culture [153, 277, 278]. Therefore, the cultivation and maintenance of AM fungal strains require different methodologies. Dalpé and Monreal [279] described the comprehensive life cycle of AM fungi and methods currently used for the propagation of inoculum and the maintenance of *in vivo* and *in vitro* source collections. As storage of spores of vesicular-arbuscular (VA) mycorrhizal fungi in soil at 5°C was satisfactory for *Glomus intraradices* but not for *Gigaspora margarita*, *Glomus mosseae*, and *Acaulospora longula*, Douds and Schenck [57] examined preservation of spores in soil at -60 to -70°C. Cryoprotectants such as DMSO, glycerol, mannitol, and sucrose were ineffective using the freeze-damage sensitive species *G. margarita*. Incubation for 47 h in 0.75 to 1.0 M trehalose conferred a measure of freeze damage protection to the spores such that germination rates of previously frozen spores of *G. margarita* were one tenth to one sixth of controls. The best method of cryoprotection and cryopreservation was found to be slow-drying of pot culture soil and freezing the spores *in situ*. This procedure was satisfactory for the five genera of VA mycorrhizal fungi evaluated. Kuszala et al. [280] demonstrated the feasibility of long-term storage of glomalean AM fungi germplasm using a simple protocol. Soil from pot cultures of the different fungal isolates was wet sieved and fractions of soil debris containing mycelium and spores were collected. These fractions were stored at room temperature (18-24 °C), +4°C, -18 °C, -80 °C, in liquid nitrogen, or were lyophilized. Sporulation was observed in all isolates stored at all temperatures. Storage in water did not affect survival of the six tested fungi. Interesting results were obtained by Declerk and Angelo-Van Coppenolle [194] in a study conducted to quantify the ability of entrapped, monoxenically produced spores of an AM fungus to germinate and reproduce the fungal life cycle after cryopreservation. No germination was obtained after incubation of entrapped spores in glycerol and mannitol and subsequent cryopreservation at -70°C, regardless of the concentration of cryoprotectants and duration of incubation. Incubation for 1 day in 0.5 M sucrose, and for 1 and 2 days in 0.5 M trehalose, led to spore germination after cryopreservation at -70°C. The highest germination rate, estimated by the percentage of potentially infective beads (%PIB), was obtained at -100°C. A %PIB of 95% (water agar medium) to 100% (Strullu-Romand medium) was obtained at this temperature. Thereafter, %PIB rapidly decreased at -140 and -180°C. Heavy sporulation and high internal root colonization were obtained after re-association of the entrapped spores, incubated for 1 d in 0.5 M trehalose and subsequently cryopreserved at -100°C, with transformed carrot roots.

Ex situ conservation of threatened terrestrial orchids requires the simultaneous conservation of their mycorrhizal associations. A method for encapsulating both seed and fungi in alginate beads (known as encapsulation-dehydration) was applied to the storage and propagation of two endangered orchid species [281]. Storage for 6 months at -18 and -196°C had no significant effect on the fungal recovery and germination percentages of either species. All beads transferred directly from *in vitro* culture to potting mix resulted in the establishment of at least one seedling, and production of a healthy tuberoid, when transferred near the commencement of the natural growing season. The implications of the long-term *ex situ* storage of orchid seed and fungal symbionts in LN for the conservation of endangered orchids (*Caladenia*, *Diuris*, *Pterostylis* and *Thelymitra*) discussed in their paper also Batty et al. [282].

Also cryopreservation of ectomycorrhizal (ECM) fungi is often difficult. Danell and Flygh [262] described a protocol for successful cryopreservation of the ectomycorrhizal mushroom *Cantharellus cibarius*, and which is very important, together with a brief summary of some unsuccessful attempts. A slow freezing rate (0.3°C per minute) of agar cubes carrying the mycelium, and a gentle addition of the cryoprotectants sorbitol and DMSO were important features. This protocol was successfully repeated by ATCC who used it to deposit a patented strain of *C. cibarius*. The viability of cryopreserved filamentous ectomycorrhizal and saprophytic fungi can be improved by a novel technique based on charcoal filter paper strips (CFS) [283]. The results indicate that axenic fungal cultures grown on CFS recovered from freezing within a few days, even though none survived cryopreservation by the conventional straw method. Fungal growth on CFS was more vigorous, with morphological differentiations such as rhizomorphs and an increased amount of aerial mycelia compared to the unamended culture media. Accordingly CFS allows the cryopreservation of a wide range of rare and important ectomycorrhizal and saprotrophic fungi, which hitherto were difficult to revive from liquid nitrogen storage with the conventional and widely applied straw technique. Obase et al. [284] determined mycelial growth and survival ratio of ECM fungi after storage at -70°C for 1, 3, or 6 months. 17 of 23 ECM fungi did not survive after storage for more than 6 months, whereas *Cenococcum geophilum*, *Lepista nuda*, and some species of *Rhizopogon* and *Suillus* did survive. A decrease in the ability of fungal isolates to form ECM associations after repeated subcultivations has been reported in several studies and has been associated with a reduced effectiveness of the fungi to improve plant growth [54, 285].

Some fungi are able to survive cryopreservation in LN without using cryoprotectants, being cooled either slowly (controlled freezing) or quickly (direct immersion in LN).

Mata et al. [286] studied five strains of edible mushrooms (*Lentinula boryana*, *Lentinula edodes*, *Pleurotus djamor*, *Pleurotus pulmonarius*, and *Volvariella volvacea*). Spawn was prepared from sorghum seeds and then incubated for 14 days under optimum conditions for each species. Once covered by mycelia, the sorghum seeds were placed in polycarbonate vials for freezing in liquid nitrogen. The effect of adding a cryoprotective solution before freezing (either 10% glycerol v/v or 5% dimethyl sulphoxide v/v) was evaluated as a function of mycelial growth and percent viability. Three main treatments were undertaken: (1) freezing with a glycerol or dimethyl sulphoxide cryoprotectant, (2) freezing with water and (3) freezing without cryoprotectant or water. Samples were maintained frozen for a week, after which time they were thawed (10 minutes at 30°C) and the seeds placed in Petri dishes with a culture medium. A recovery rate of 96.8% was obtained for the total number of samples summed over all strains and treatments. In contrast, 99.2% of the samples frozen without cryoprotectant were recovered. The recovery of frozen mycelia was delayed with respect to a control group, which was not frozen. However, no difference was observed in percent recovery and mycelial diameter when a new series of spawn was prepared from mycelia that had been previously frozen. The results obtained from this experiment demonstrate that an adequate recovery of mycelia can be obtained without using a cryoprotectant. This capacity might enable large quantities of commercial mushroom strains to be handled at reduced production costs. It is suggested that the mycelia survived freezing without cryoprotectants because they were embedded and protected within the sorghum seeds used to elaborate the spawn. In our laboratory we also tested the possibility to cryopreserve the cultures in the absence of a cryoprotectant on a set of 60 basidiomycete strains (by direct immersion into LN or by controlled freezing -1 °C per minute), but the results were not very encouraging - only about 10% of cultures survived after 48 hours [Homolka, unpublished results].

Mycelial stock cultures of *Agaricus bisporus*, *A. bitorquis*, *Pleurotus flabellatus*, *P. sajor-caju*, *P. ostreatus*, *P. sapidus*, *Auricularia polytricha*, *Lentinula edodes*, *Morchella esculenta* and *Volvariella volvacea* were maintained by frequent subculturing at an interval of two months and separately as wheat grain spawn in LN with 15% glycerol [19]. The latter method turned out to be better for the maintenance. The percent recoveries of stored samples were unchanged from the first recovery after six months to the last recovery after 42 months in 9 out of 11 stock cultures preserved under LN. However, a marginal decline in survival of 10 % was recorded in *Auricularia polytricha* and *Volvariella volvacea*. Yields before preservation of mushroom stock cultures and after 30 months of preservation exhibited static biological efficiency and fruitbody weight. The comparison of Random Amplified Polymorphic DNA (RAPD) and Internal Transcribed Spacers (ITS) PCR amplified products did not exhibit DNA fragment variation in banding patterns at the intraspecific level during preservation of stock cultures by either method. Spawn stock of the cultivated mushroom was stable at least 26 months after LN refrigeration [287].

Paleontological studies of *naturally cryopreserved organisms* brought interesting results. Almost all of the investigated samples of the Arctic and Antarctic permafrost sediments of different genesis with ages from 5-10 thousand to 2-3 million years were found to contain viable micromycete and bacterial cells. The maximum amounts of viable cells of fungi (up to 10^4 CFU/g air-dried sample) and bacteria (up to 10^7-10^9 CFU/g air-dried sample) were present in fine peaty sediment samples taken from different depths. The identified micromycetes belonged to more than 20 genera of the divisions *Basidiomycota*, *Ascomycota*, and *Zygomycota*, and some represented mitosporic fungi. Thawing the samples at 35 and 52°C allowed the number of detected fungal genera to be increased by more than 30%. The analysis of the isolated fungi showed that most of them originated from the microbial communities of ancient terrestrial biocenoses [288].

Conclusion

The primary methods of culture preservation are continuous growth, drying, and freezing. A traditional method of serial transfers, which is still frequently (sometimes indispensably) used in many laboratories, is far from being optimal and does not prevent genetic and physiological changes during long-term maintenance. Various storage methods have been developed during the time. Among them, besides lyophilization, the storage in liquid nitrogen has been considered the best and most widely applicable preservation technique available for filamentous fungi. It is a safe and reliable method of a long-term maintenance of most fungal species, especially those not amenable to freeze-drying, and it can be used for both sporulating and non-sporulating (mycelial) fungal cultures. In our opinion it is highly desirable to extend the current knowledge of the effects of cryopreservation on enzyme and other metabolite production to a wider set of biotechnologically-relevant enzymes and metabolites and to provide detailed information on the genetic stability of the cryopreserved material.

Acknowledgments

I thank Drs Zdeněk Hubálek, David Smith, Matthew J. Ryan, Joost A. Stalpers, Agnes H. S. Onions, Karen K. Nakasone and her colleagues, from whose work I have drawn inspiration.

References

[1] Smith, D The use of cryopreservation in the ex-situ conservation of fungi. *CryoLetters* 1998; 19: 79-90.

[2] Prescott, JM; Kernkamp, MF. Genetic stability of *Puccinia graminis tritici* in cryogenic storage. *Plant Dis. Rep.* 1971; 55: 695–696.

[3] Heckly, RJ. Preservation of microorganisms. *Adv. Appl. Microbiol.* 1978; 24: 1-53.

[4] Smith, D. Liquid nitrogen storage of fungi. *Trans. Br. Mycol. Soc.* 1982; 79: 415– 421.

[5] Ryan, MJ; Smith, D; Jeffries, P. A decision-based key to determine the most appropriate protocol for the preservation of fungi. *World J. Microbiol. Biotechnol.* 2000; 16: 183-186.

[6] Smith, D; Ryan, MJ. The impact of OECD best practice on the validation of cryopreservation techniques for microorganisms. *CryoLetters* 2008; 29: 63-72.

[7] Smith, D; Ryan, MJ. Implementing best practices and validation of cryopreservation techniques for microorganisms. *The ScientificWorld Journal*, Vol. 2012, Article ID 805659, doi:10.1100/2012/805659.

[8] Wiest, A; Schnittker, R; Plamann, M; McCluskey, K. Best practices for fungal germplasm repositories and perspectives on their implementation. *Appl. Microbiol. Biotechnol.* 2012; 93: 975-982.

[9] Smith, D. Provision and maintenance of micro-organisms for industry and international research networks. *CryoLetters* 2001; 22: 91-96.

[10] Stacey, GN; Day, JG. Ex-situ conservation of biological resources and the role of biological resource centers. In: Day, JG, Stacey, GN (Eds). *Methods in Molecular Biology 368: Cryopreservation and freeze-drying protocols.* 2nd edn. Totowa: Humana Press; 2007; 1-14.

[11] Hubálek, Z. *Cryopreservation of microorganisms at ultra-low temperatures.* Prague: Academia Publishing House; 1996.

[12] Homolka, L. On the problem of maintenance and cultivation of higher fungi. *Folia Microbiol.* 1976; 21: 189-190.

[13] Stoychev, I; Homolka, L.; Nerud, F; Lisá, L. Activities of ligninolytic enzymes in some white-rot basidiomycete strains after recovering from cryopreservation in liquid nitrogen. *Antonie van Leeuwenhoek* 1998; 73: 211-214.

[14] Homolka, L.; Lisá, L.; Nerud, F. Basidiomycete cryopreservation on perlite: evaluation of a new method. *Cryobiology* 2006; 52: 446-453.

[15] Ryan, MJ; Smith, D; Bridge, PD; Jeffries, P. The relationship between fungal preservation method and secondary metabolite production in *Metarhizium anisopliae* and *Fusarium oxysporum*. *World J. Microbiol. Biotechnol.* 2003; 19: 839-844.

[16] Sundari, SK; Adholeya, A. Functional integrity and structural stability of freeze- dried ectomycorrhizal fungi established through viability assays. *Can. J.Microbiol.* 2001; 47: 172-177.

[17] Smith, D. Tolerance to freezing and thawing. In: Jennings, DH (Ed.). *Stress Tolerance of Fungi.* New York: Marcel Dekker; 1993; 145–171.

[18] Dahmen, H; Staub, TH; Schwinn, FJ. Technique for long-term preservation of phytopathogenic fungi in liquid nitrogen. *Phytopathology* 1983; 73: 241-246.
[19] Singh, SK; Upadhyay, RC; Kamal, S; Tiwari, M. Mushroom cryopreservation and its effect on survival, yield and genetic stability. *CryoLetters* 2004; 25: 23-32.
[20] Ryan, MJ; Jeffries, P; Bridge, PD; Smith, D. Developing cryopreservation protocols to secure fungal gene function. *CryoLetters* 2001; 22: 115-124.
[21] Stummer, BE; Zanker, T; Scott, ES. Cryopreservation of air-dried conidia of *Uncinula necator*. *Australasian Plant Pathol.* 1999; 28: 82-84.
[22] Voyron, S; Roussel, S; Munaut, F; Varese, GC; Ginepro, M; Declerck, S; Filipello Marchisio, V. Vitality and genetic fidelity of white-rot fungi mycelia following different methods of preservation. *Mycol. Res.* 2009; 113: 1027-1038.
[23] Camelini, CM; Pena, DA; Gomes, A; Steindel, M; Rossi, MJ; Giachini, AJ; de Mendonca, MM. An efficient technique for in vitro preservation of *Agaricus subrufescens* (=*A. brasiliensis*). *Ann.Microbiol.* 2012; 62: 1279-1285.
[24] Homolka, L; Lisá, L; Eichlerová, I; Valášková, V; Baldrian, P.. Effect of long-term preservation of basidiomycetes on perlite in liquid nitrogen on their growth, morphological, enzymatic and genetic characteristics. *Fungal Biology* 2010; 114: 929- 935.
[25] Arabi, MIE; Jawhar, M; Al-Daoude, A. Viability of *Cochliobolus sativus* cultures after storage under different conditions. *J. Plant Pathol.* 2007; 89: 79-83.
[26] Broughton, R; Buddie, AG; Smith, D; Ryan, MJ. The effect of cryopreservation on genomic stability in strains of the fungus *Trichoderma*. *CryoLetters* 2012; 33: 299- 306.
[27] Ryan, MJ; Smith, D. Fungal genetic resource centres and the genomic challenge. *Mycol. Res.* 2004; 108: 1351-1362.
[28] Abd-Elsalam, KA; Yassin, MA; Moslem, MA; Bahkali, AH; de Wit, PJGM; McKenzie, EHC; Stephenson, SL; Cai, L; Hyde, KD. Culture collections, the new herbaria for fungal pathogens. *Fungal Diversity* 2010; 45: 21-32.
[29] Tedeschi, R; De Paoli, P. Collection and preservation of frozen microorganisms. Methods in biobanking. In: *Dillner, J (Ed.) Methods in Molecular Biology*. New York: Humana Press; 2011; 675: 313-326.
[30] Costa, CP; Ferreira, MC. Preservation of microorganisms - a review. *Rev. Microbiol.* 1991; 22: 263-268.
[31] Tan, CS. Preservation of fungi. *Cryptogamie Mycologie* 1997; 18: 157-163.

[32] Nakasone, KK; Peterson, SW; Jong, SC. Preservation and distribution of fungal cultures. In: Mueller, GM; Bills, GF; Foster, MS (eds.) *Biodiversity of fungi. Inventory and monitoring methods.* Amsterdam: Elsevier Academic Press; 2004; 35- 47.

[33] Nagai, T; Tomioka, K; Takeuchi, K; Iida, M; Kawada, M; Sato, T. Evaluation of preservation techniques of microorganism resources in the MAFF Genebank. *JARQ- Jap. Agr. Res. Quart.* 2005; 39: 19-27.

[34] Meryman, H.T., 1966. Review of biological freezing. In: Meryman, HT (ed.) *Cryobiology.* London and New York: Academic Press; 1966; 1-114.

[35] Mazur, P. Physical and chemical basis of injury in single-celled microorganisms subjected to freezing and thawing. In: Meryman, HT (ed.) *Cryobiology.* London and New York: Academic Press 1966; 213-315.

[36] Mazur. P. Freezing of living cells: mechanisms and implications. *Am. J. Physiol.* 247 (*Cell Physiol.* 16) 1984; C125-C142.

[37] Mazur, P. 2004. Principles of cryobiology. In: Fuller, BJ; Lane, N; Benson, E (eds) *Life in the Frozen State.* Boca Raton: CRC Press 2004; 3-65.

[38] Bailey, CA (ed.) *Advanced cryogenics.* New York and London: Plenum Press; 1971; 527 pp.

[39] Haselden, GG (ed.) *Cryogenic fundamentals.* London: Academic Press; 1971.

[40] Williams, M. *A comprehensive guide to deep freezing.* Feltham: P. Hamlyn; 1973; 94 pp.

[41] Grout, BWW, Morris, GJ (eds.) *The effects of low temperatures on biological systems.* London: E. Arnold; 1987; 500 pp.

[42] Hubálek, Z. Protectants used in the cryopreservation of microorganisms. *Cryobiology* 2003; 46: 205-229.

[43] Smith, D. Cryoprotectants and the cryopreservation of fungi. *Trans. Br. Mycol. Soc.* 1983; 80: 360-363.

[44] Day, JG; Stacey, GN (eds.) *Methods in Molecular Biology, vol. 368: Cryopreservation and freeze-drying protocols.* 2nd edition. Totowa, NJ: Humana Press; 2007; 348pp.

[45] Lopez, E; Cipri, K; Naso, V. Technologies for cryopreservation: Overview and innovation. In: Katkov, I.I. (ed.) *Current Frontiers in Cryobiology*; InTech; 2012; 527- 546. Available for free from: http:// www.intechopen.com/books/current-frontiers-in- cryobiology/technologies-for-cryopreservation-overview-and-innovation.

[46] Franks, F. Biophysics and biochemistry of low temperatures and freezing. In: Morris, G J; Clarke, A.(eds.) *Effects of Low Temperature of Biological Membranes*; London, UK: Academic Press; 1981; 3–19.

[47] Muldrew, K; McGan, LE. *Cryobiology - A short course*; 1999; Free download at http://people.ucalgary.ca/~kmuldrew/cryo_course/course_outline.html.
[48] Zhmakin, A.I. *Fundamentals of cryobiology. Physical phenomena and mathematical models*. 2009; XIV; 278 p.
[49] Katkov, I.I. (ed.) *Current frontiers in cryobiology*. Rijeka, Croatia: InTech; 2012.
[50] Križková, L; Balan, J. Prolonged preservation of *Oomycetes* and predacious *Fungi imperfecti* under paraffin oil. *Folia Microbiol.* 1975; 20: 351-353.
[51] Perrin, PW. Long-term storage of cultures of wood-inhabiting fungi under mineral oil. *Mycologia* 1979; 71: 867-869.
[52] Johnson, GC; Martin, AK. Survival of wood-inhabiting fungi stored for 10 years in water and under oil. *Can. J. Microbiol.* 1992; 38: 861-864.
[53] Castellani, A. Further researches on the long viability and growth of many pathogenic fungi and some bacteria in sterile distilled water. *Mycopathol. Mycol. Appl.* 1963; 20: 1-6.
[54] Marx, DH; Daniel, WJ. Maintaining cultures of ectomycorrhizal and plant pathogenic fungi in sterile water cold storage. *Can. J. Microbiol.* 1976; 22: 338-341.
[55] Ellis, JJ. Preserving fungus strains in sterile water. *Mycologia* 1979; 71: 1072-1075.
[56] Richter, DL; Bruhn, JN. Revival of saprotrophic and mycorrhizal basidiomycete cultures from cold storage in sterile water. *Can. J. Microbiol.* 1989; 35: 1055-1060.
[57] Douds, DD, Jr.; Schenck, NC. Cryopreservation of spores of vesicular-arbuscular mycorrhizal fungi. *New Phytologist* 1990; 115: 667-674.
[58] *Preservation methods: freezing and freeze-drying*; 2nd edition. Rockville, Md.: American Type Culture Collection; 1991.
[59] Greuter, W; McNeill, J; Barrie, FR; Burdet, HM; Demoulin, V; Filgueiras, TS; Nicolson, DH; Silva, PC; Skog, J; Trehane, P; Turland, N J; Hawksworth, DL.: *International code of botanical nomenclature (Saint Louis Code) adopted by the Sixteenth International Botanical Congress, St Louis, Missouri, July-August 1999*; Regnum Veg. 138; 2000; xviii + 474 pp.
[60] Smith, D; Onions, AHS. A comparison of some preservation techniques for fungi. *Trans. Br. Mycol. Soc.* 1983; 81: 535–540.
[61] Homolka, L; Lisá, L; Eichlerová, I; Nerud, F. Cryopreservation of some basidiomycete strains using perlite. *J. Microbiol. Methods* 2001; 47: 307-313.
[62] Homolka, L; Lisá, L; Kubátová, A; Váňová, M; Janderová, B; Nerud, F. Cryopreservation of filamentous micromycetes and yeasts using perlite. *Folia Microbiol.* 2007; 52: 153-157.

[63] Smith, D; Onions, AHS. *The preservation and maintenance of living fungi.* 2nd Edition, Egham (UK): International Mycological Institute, CAB International; IMI Technical Handbooks No. 2; 1994.

[64] Jong, SC; Davis, EE. Germplasm preservation of edible fungi in culture through cryogenic storage, In: *Scientific and Technical Aspects of Cultivating Edible Fungi.* Proc. Internat. Symp.; The Penna State University, University Park (USA); 1986; 213- 225.

[65] Onions, AHS. Preservation of fungi. In: Booth, C (ed.); *Methods in Microbiology*, Vol. 4, 1971; New York and London: Academic Press; 113-115.

[66] Kuek C., Issues concerning the production and use of inocula of ectomycorrhizal fungi in increasing the economic productivity of plantations. In: Robson, AD; Abbott, LK; Malajczuk, N (eds). *Management of mycorrhizas in agriculture, horticulture and forestry.* Dordrecht: Kluwer Academic Publishers; 1994; 221-230.

[67] Corbery, Y; Le Tacon, F. Storage of ectomycorrhizal fungi by freezing. *Ann. Sci. forest.* 1997; 54: 211-217.

[68] Siddiqui, ZA, Kataoka, R. Mycorrhizal inoculants: progress in inoculant production technology. In: Ahmad I; Ahmad F; Pitchel J (eds); *Microbes and Microbial Technology: Agricultural and Environmental Applications.* New York: Springer; 2011; 489-506.

[69] Crespo MJ; Abarca ML; Cabañes FJ. Evaluation of different preservation and storage methods for *Malassezia* spp. *J. Clin. Microbiol.* 2000; 38: 3872-3875.

[70] Lopez Lastra, CC; Hajek, EA; Humber, RA. Comparing methods of preservation for cultures of entomopathogenic fungi. *Can. J. Bot.* 2002; 80: 1126-1130.

[71] Homolka, L; Lisá, L. Long-term maintenance of fungal cultures on perlite in cryovials - an alternative for agar slants. *Folia Microbiol.* 2008; 53: 534-536.

[72] Tan, CS; Stalpers, JA; van Ingen, CW. Freeze-drying of fungal hyphae. *Mycologia* 1991; 83: 654–657.

[73] Tan, CS; van Ingen, CW; Stalpers, JA. Freeze-drying of fungal hyphae and stability of the product. In: A. van Griensven (ed.); *Genetics and breeding of Agarics*; Wageningen, The Netherlands: Pudoc; 1991.

[74] Fortney, KF; Thoma, RW. Stabilization of culture productivity. *Dev. Ind. Microbiol.* 1977; 18: 319-325.

[75] Burdsall, HH; Dorworth, EB. Preserving cultures of wood-decaying *Basidiomycotina* using sterile distilled water in cryovials. *Mycologia* 1994; 86: 275-280.

[76] Onions, AHS. Storage of fungi by mineral oil and silica gel for use in collections with limited resources. In: Pestana de Castro, AF; Da Silva, EJ; Skerman, VBD Leveritt, WW (eds); *Proceedings of the second international conference on culture collections,* University of Queensland, Brisbane; Brussels: World Federation for Culture Collections 1977; 104-113.

[77] Kochetkov, VV. Variability of fungi cultures after 25-30 year storage under vaseline oil; In: Samson, RA; Stalpers, JA; van der Mei, D; Stouthamer A.H. (Eds). *Culture collections to improve the quality of life. Proceedings of the Eight International Congress for Culture Collections.* Veldhoven, The Netherlands, 25-29 August 1996. Baarn: Centraalbureau voor Schimmelcultures, Baarn, The Netherlands and the World Federation for Culture Collections; 1996; 446-447.

[78] Piaggio, MJ. Survival of fungal strains stored for 20 years under oil. In: Samson, RA; Stalpers, JA; van der Mei, D; Stouthamer A.H. (Eds). *Culture collections to improve the quality of life. Proceedings of the Eight International Congress for Culture Collections.* Veldhoven, The Netherlands, 25-29 August 1996. Baarn: Centraalbureau voor Schimmelcultures, Baarn, The Netherlands and the World Federation for Culture Collections; 1996; 444-445.

[79] Castellani, A. The viability of some pathogenic fungi in sterile distilled water. *J. Trop. Med. Hyg.* 1939; 42: 225-226.

[80] Castellani, A. Maintenance and cultivation of common pathogenic fungi of man in sterile distilled water, further researches. *J. Trop. Med. Hyg.* 1967; 70: 181-184.

[81] Clark, G; Dick, MW. Long-term storage and viability of aquatic oomycetes. *Trans. Br. Mycol. Soc.* 1974; 63: 611-612.

[82] Croan, SC; Burdsall, HH, Jr; Rentmeester, RM. Preservation of tropical wood- inhabiting basidiomycetes. *Mycologia* 1999; 91: 908-916.

[83] Smith, JE; McKay, D; Molina, R. Survival of mycorrhizal fungal isolates stored in sterile water at two temperatures and retrieved on solid and liquid nutrient media. *Can. J. Microbiol.* 1994; 40: 736-742.

[84] Boeswinkel, HJ. Storage of fungal cultures in water. *Trans. Br. Mycol. Soc.* 1976; 66: 183-185.

[85] McGinnis, M R; Padhye, AA; Ajello, L. Storage of stock cultures of filamentous fungi, yeasts, and some aerobic actinomycetes in sterile distilled water. *Appl. Microbiol.* 1974; 28:218-222.

[86] Onions, AHS; Smith, D. Current status of culture preservation and technology. In: Batra, LR; Iigima, T (eds). *Critical problems of culture collections.* Osaka: Institute of Fermentation 1984; 41-47.

[87] Hartung de Capriles, C; Mata, S; Middelveen, M. Preservation of fungi in water (Castellani): 20 years. *Mycopathologia* 1989; 106: 73-80.

[88] Qiangqiang, Z; Jiajun, W; Li, L. Storage of fungi using sterile distilled water or lyophilization: comparison after 12 years. *Mycoses* 1998; 41: 255-257.

[89] Hajek, AE; Humber, RA; Griggs MH. Decline in virulence of *Entomophaga maimaiga* (*Zygomycetes*: *Entomophthorales*) with repeated in vitro subculture. *J Invertebr. Pathol.* 1990; 56:91-97.

[90] Hajek, AE; Shimazu, M; Humber, RA. Instability in pathogenicity of *Entomophaga maimaiga* after long-term cryopreservation. *Mycologia* 1995; 87:483-489.

[91] Samson, RA; McCoy, C W; O'Donnell, KL.. Taxonomy of the acarine parasite *Hirsutella thompsonii*. *Mycologia* 1980; 72: 359-377.

[92] Kumar, PS; Singh, L. A simple method of storing nonsynnematous and synnematous isolates of *Hirsutella thompsonii* while conserving their pleomorphism, pathogenicity and genetic purity. *Folia Microbiol.* 2012; 57:15-19.

[93] Borman, AM; Szekely, A; Campbell, CK; Johnson, EM. Evaluation of the viability of pathogenic filamentous fungi after prolonged storage in sterile water and review of recent published studies on storage methods. *Mycopathologia* 2006; 161: 361-368.

[94] Richter, DL; Kangas, LC; Smith, JK; Laks, PE. Comparison of effectiveness of wood decay fungi maintained by annual subculture on agar and stored in sterile water for 18 years. *Can. J. Microbiol.* 2010; 56: 268-271.

[95] Richter, DL. Revival of saprotrophic and mycorrhizal basidiomycete cultures after 20 years in cold storage in sterile water. *Can. J. Microbiol.* 2008; 54: 595-599.

[96] Deshmukh, SK. The maintenance and preservation of keratinophilic fungi and related dermatophytes. *Mycoses* 2003; 46: 203-207.

[97] Sharma, B; Smith, D. Recovery of fungi after storage for over a quarter of a century. *World J. Microbiol. Biotechnol.* 1999; 15: 517-519.

[98] Horaczek, A; Viernstein, H. Comparison of three commonly used drying technologies with respect to activity and longevity of aerial conidia of *Beauveria brongniartii* and *Metarhizium anisopliae*. *Biol. Control* 2004; 31: 65-71.

[99] Quintanilla, LG; Amigo, J; Pangua, E; Pajaron, S. Effect of storage method on spore viability in five globally threatened fern species. *Ann. Bot. (London)* 2002; 90: 461- 467.

[100] Delatour, C. A very simple method for long-term storage of fungal cultures. *Eur. J. Forest Pathol.* 1991; 21: 444–445.

[101] Sneh, B; Burpee, L; Ogoshi, A. *Identification of Rhizoctonia species*; St. Paul Minnesota: The American Phytopathological Society Press 1991; 133 pp.
[102] Singleton, LL; Mihail, JD; Rush, CM. *Methods for research on soilborn phytopathogenic fungi*; St. Paul Minnesota: The American Phytopathological Society Press 1992; 115-128 and 157-165.
[103] Nuzum, C. A simple method for the preservation of some non-sporing fungi. *Australasian Plant Pathol.* 1989; 18: 104-105.
[104] Steinkraus, DC; Boys, GO; Slaymaker, PH. Culture, storage, and incubation period of *Neozygites fresenii* (*Entomophthorales*: *Neozygitaceae*) a pathogen of the cotton aphid. *Southwestern Entomologist* 1993; 18: 197-202.
[105] Shearer, BL; Zeyen, RJ; Ooka, JJ. Storage and behaviour in soil of *Septoria* species isolated from cereals. *Phytopathology* 1974; 64: 163-167.
[106] Reinecke, P; Fokema, NJ.. *Pseudocercosporella herpotrichoides*: storage and mass production of conidia. *Trans. Br. Mycol. Soc.* 1979; 72: 329-331.
[107] Perkins, DD. Preservation of *Neurospora* stock cultures with anhydrous silica gel. *Can. J. Microbiol.* 1962; 8: 591-594.
[108] Raper, KB. *The Dictyostelids*; Princeton, New Jersey: Princeton University Press; 1984.
[109] Jong, SC; Birmingham, JM. Cultivation and preservation of fungi in culture. In: McLaughlin, DJ; McLaughlin, EG; Lemke, PA (eds.); *The Mycota,* part B; Berlin: Springer Verlag; 2001; 193-202.
[110] Tan, CS. Preservation of fungi. *Cryptogamie Mycologie* 1997; 18: 157–163.
[111] Franks, F. Freeze drying: from empiricism to predictability. *CryoLetters* 1990; 11: 93-110.
[112] Tan, CS; van Ingen, CW; Stalpers, JA. Freeze-drying fungi using a shelf freeze-drier; In: Day, JG; Stacey, GN (eds.); *Methods in Molecular Biology, vol. 368: Cryopreservation and Freeze-Drying Protocols, 2nd Edition*; Totowa, NJ: Humana Press Inc.; 2007; 119-126.
[113] Raper, KB; Alexander, DF. Preservation of molds by lyophil process. *Mycologia* 1945; 37: 499-525.
[114] Bunse, T; Steigleder, GK. The preservation of fungal cultures by lyophilization. *Mycoses* 1991; 34: 173-176.
[115] Croan, SC. *Lyophilization of hypha-forming tropical wood-inhabiting Basidiomycotina. Mycologia* 2000; 92: 810-817.
[116] Pertot, E; Puc, A; Kremser, M. Lyophilization of nonsporulating strains of the fungus *Claviceps. Eur. J. Appl.Microbiol.* 1977; 4: 289-294.
[117] Ryan MJ, Smith D, 2007. Cryopreservation and freeze-drying of fungi employing centrifugal and shelf freeze-drying. pp. In: Day, JG; Stacey, GN (eds.); *Methods in Molecular Biology*, vol. 368: *Cryopreservation and*

Freeze-Drying Protocols, 2nd Edition; Totowa, NJ: Humana Press Inc.; 2007; 127-140.

[118] Sundari, SK; Adholeya, A. Freeze-drying vegetative mycelium of *Laccaria fraterna* and its subsequent regeneration. *Biotechnol. Techniques* 1999; 13: 491-495.

[119] Sundari, SK; Adholeya, A. Retention of enzyme activity following freeze-drying the mycelium of ectomycorrhizal isolates. *World J. Microbiol. Biotechnol.* 2000; 16: 373- 376.

[120] Carvalho, RO; Braga, FR; Araujo, JV. Viability and nematophagous activity of the freeze-dried fungus *Arthrobotrys robus*ta against *Ancylostoma* spp. infective larvae in dogs. *Vet. Parasitol.* 2011; 176: 236-239.

[121] Toegel, S; Salar-Behzadi, S; Horaczek-Clausen, A; Viernstein, H. Preservation of aerial conidia and biomasses from entomopathogenic fungi *Beauveria brongniartii* and *Metarhizium anisopliae* during lyophilization. *J. Invertebr. Pathol.* 2010; 105: 16- 23.

[122] Rybnikář, A. Long-term maintenance of lyophilized fungal cultures of the genera *Epidermophyton*, *Microsporum*, *Paecilomyces* and *Trichophyton*. *Mycoses* 1995; 38: 145-147.

[123] Schoenleincrusius, IH; Okino, LK; Lucon, CMM. Survival of fungi preserved by lyophilization after 49 years. *Rev. Microbiol.* 1993; 24: 207-211.

[124] Day, JG; Harding, K; Nadarajan, J; Benson, EE. Cryopreservation: conservation of bioresources at ultra-low temperatures. In: Walker JM; Rapley, R (eds); *Molecular Biomethods Handbook,* 2nd edn.; Totowa: Humana Press; 2008; 917-947.

[125] Tan, CS; Vlug IJA; Stalpers, JA; van Ingen, CW. Microscopical observations on the influence of the cooling rate during freeze-drying of conidia. *Mycologia* 1994; 86: 281-289.

[126] Carmichael, JW. Frozen storage for stock cultures of fungi. *Mycologia* 1956; 48: 378-381.

[127] Carmichael, JW. Viability of mold cultures stored at -20°C. *Mycologia* 1962; 54: 432- 436.

[128] Pasarell, L; McGinnis, MR. Viability of fungal cultures maintained at -70°C. *J. Clin. Microbiol.* 1992; 30: 1000-1004.

[129] Ito, T; Yokoyama, T. Preservation of basidiomycete cultures by freezing. *IFO Res. Commun.* 1983; 11: 60-70.

[130] Ito, T; Yokoyama, T. Further investigation on the preservation of basidiomycete cultures by freezing. *IFO Res. Commun.* 1987; 13: 69-81.

[131] Ito, T. 1991. Frozen storage of fungal cultures deposited in the IFO Culture Collection. IFO Res. Commun. 15: 119-128.

[132] Ito, T; Nakagiri, A. Viability of frozen cultures of basidiomycetes after fifteen-year storage. Microbiol. Cult. Coll. 1996; 12: 67-78.

[133] Chandler, D. Cryopreservation of fungal spores using porous beads. *Mycol. Res*. 1994; 98: 525-526.

[134] Belkacemi, L; Barton, RC; Evans, EGV. Cryopreservation of *Aspergillus fumigatus* stock cultures with a commercial bead system. *Mycoses* 1997; 40: 103-104.

[135] Palágyi, Z; Nagy, Á; Vastag, M; Ferency, L; Vágvölgyi, C. Maintenance of fungal strains on cryopreservative-immersed porous ceramic beads. *Biotechnol.Tech*. 1997; 11: 249-250.

[136] Morris, GJ. *Cryopreservation: An introduction to cryopreservation in culture collections.* Cambridge, UK: Culture Centre of Algae and Protozoa. ITE; 1981.

[137] Kirsop, B; Snell, JJS (eds.) *Maintenance of microorganisms*. London, New York: Academic Press; 1984.

[138] Kirsop, BE; Doyle, A. *Maintenance of microorganisms and cultured cells: A manual of laboratory methods*. London, UK: Academic Press; 1991.

[139] Elliott, TJ. Alternative ampoule for storing fungal cultures in liquid nitrogen. *Trans. Br. Mycol. Soc*. 1976; 67: 545-546.

[140] Stalpers, JA; de Hoog, A; Vlug, IJ. Improvement of the straw technique for the preservation of fungi in liquid nitrogen. *Mycologia* 1987; 79: 82–89.

[141] Smith, D; Kolkowski, J. *Fungi. Preservation and Maintenance of cultures used in Biotechnology and Industry*. Stoneham, MA: Butterworth Publishing Company; 1992.

[142] Kolkowski, J.A., Smith, D.: Cryopreservation and freeze-drying of fungi. In: Day, J0; McLellan, MR (eds); *Methods in Molecular Biology*, Vol. 38: *Cryopreservation and freeze-drying protocols*; Totowa, NJ: Humana Press Inc.; 1995; 49-61.

[143] Hwang, SW. Effects of ultra-low temperatures on the viability of selected fungus strains. *Mycologia* 1960; 52: 527-529.

[144] Hwang, SW. Long term preservation of fungus cultures with liquid nitrogen refrigeration. *Appl. Microbiol*. 1966; 14: 784–788.

[145] Butterfield, W; Jong, S; Alexander, MT. Preservation of living fungi pathogenic for man and animals. *Can. J. Microbiol*. 1974; 20: 1665-1673.

[146] Greiff, D; Melton, H; Rowe, TW. On the sealing of gas-filled glass ampoules. *Cryobiology* 1975; 12: 1-14.

[147] Butterfield, W; Jong, S; Alexander, MT. Polypropylene vials for preserving fungi in liquid nitrogen. *Mycologia* 1978; 70: 1122-1124.

[148] Simione, FP; Daggett, PM; McGrath, MS; Alexander, MT. The use of plastic ampoules for freeze preservation of microorganisms. *Cryobiology* 1977; 14: 500-502.

[149] Dietz, A. Nitrogen preservation of stock cultures of unicellular and filamentous microogranisms. In: *The cryogenic preservation of cell cultures*; Rinfret, A.P., Lasalle, B. (eds); Washington: Nat. Acad. Sci.; 1975; 22-36.

[150] Elliott, TJ; Challen, MP. *The storage of mushroom strains in liquid nitrogen*; Littlehampton, W. Sussex: *The Glasshouse Crops Res. Inst. Ann. Rep.* 1979; 194-204.

[151] Challen, MP; Elliott, T. Polypropylene straw ampoules for the storage of microorganisms in liquid nitrogen. *J. Microbiol. Methods* 1986; 5: 11-23.

[152] Hoffmann, P. Cryopreservation of fungi. *World J. Microbiol. Biotechnol.* 1991; 7: 92- 94.

[153] Homolka, L; Lisá, L; Nerud, F. Viability of basidiomycete strains after cryopreservation: Comparison of two different freezing protocols. *Folia Microbiol.* 2003; 48: 219-226.

[154] Thammavongs, B; Poncet, JM; Desmasures, N; Gueguen, M; Panoff, JM. Resin straw as an alternative system to securely store frozen microorganisms. *J. Microbiol. Methods* 2004; 57: 181-186.

[155] Hwang, SW. Investigation of ultra low temperature for fungal cultures I. An evaluation of liquid nitrogen storage for preservation of selected fungal cultures. *Mycologia* 1968; 60: 613-621.

[156] Tuite, J. Liquid nitrogen storage of fungi sealed in a polyester film. *Mycologia* 1968; 60: 591-594.

[157] Feltham, RKA; Power, AK; Pell, PA; Sneath, PHA. A simple method for storage of bacteria at -76° *C. J. Appl. Bacteriol.* 1978; 44: 313-316.

[158] Fuller, BJ. Cryoprotectants: the essential antifreezes to protect life in the frozen state. *CryoLetters* 2004; 25: 375-388.

[159] Stout, D. [First English Translation of] 'Untersuchungen über Das Erfieren der Pflanzen' (Investigations into the freezing of plants); [Monograph by H. Molisch, 1897.] *CryoLetters* 1982; 3: 331-390.

[160] Maximow, NA. Chemische Schutzmittel der Pflanzen gegen Erfrieren. *Ber. Deutsch. bot. Gess.* 1912; 30: 52-65.

[161] Polge, C; Smith, AU; Parkes, S. Revival of spermatozoa after dehydration at low temperatures. *Nature* 1949; 164: 666.

[162] Tan, CS; van Ingen, CW; Talsma, H; van Miltenburg, JC; Steffensen, CL; Vlug, IJA; Stalpers, JA. Freeze-drying of fungi: Influence of composition

and glass transition temperature of the protectant. *Cryobiology* 1995; 32: 60-67.

[163] Roquebert, MF; Bury, E. Effect of freezing and thawing on cell membranes of *Lentinus edodes*, the Shiitake mushroom. *World J. Microbiol. Biotechnol.* 1993; 9: 641-647.

[164] Mata, G; Salmones, D; Pérez, R; Guzmán, G. Behavior of some strains of the genus *Pleurotus* after different procedures for freezing in liquid nitrogen. *Rev.Microbiol. Sao Paulo* 1994; 25: 197-200.

[165] Ohmasa, M; Abe, Y; Babasaki, K; Hiraide, M; Okabe, K. Preservation of cultures of mushrooms by freezing. *Trans Mycol. Soc. Jap.* 1992; 33: 467-479.

[166] Sakurada, M; Tsuzuki, Y; Morgavi, DP; Tomita, Y; Onodera, R. Simple method for cryopreservation of an anaerobic rumen fungus using ethylene-glycol and rumen fluid. *FEMS Microbiol. Lett.* 1995; 127: 171-174.

[167] Griffith, M; Ewart, KV. Antifreeze proteins and their potential use in frozen foods. *Biotechnol. Adv.* 1995; 13: 375-402.

[168] Leef, J; Mazur, P. Physiological response of *Neurospora* conidia to freezing in the dehydrated, hydrated or germinated state. *Appl. Environ. Microbiol.* 1978; 35: 72-83.

[169] Anon. *Laboratory procedures for microorganisms. M/1998/3.00 Appendix 5.07, Cryopreservation of fungi. Common Access to Biological Resources and Information (CABRI)*, http://www.cabri.org/; 1998. Ryan, MJ; Jeffries, P; Bridge, PD; Smith, D. Developing cryopreservation protocols to secure gene function. *CryoLetters* 2001; 22: 115-124.

[170] Wellman, AM; Walden, DB. Qualitative and quantitative estimates of viability for some fungi after periods of storage in liquid nitrogen. *Can. J. Microbiol.* 1964; 10: 585-593.

[171] Goos, RD; Davis, EE; Butterfield, W. Effect of warming rates on the viability of frozen fungous spores. *Mycologia* 1967; 59: 58-66.

[172] Hwang, SW; Howell, A. Investigation of ultra-low temperature of fungal cultures II. Cryoprotection afforded by glycerol and dimethyl sulphoxide to eight selected fungal cultures. *Mycologia* 1968; 60: 622–626.

[173] Mazur, P. The effects of subzero temperatures on micro-organisms. In: Parkes, AS; Smith, AU (eds.); *Recent research in freezing and drying*; Oxford: Blackwell; 1960. 65-77.

[174] Mazur, P. Physical factors implicated in the death of microorganisms at subzero temperatures. *Ann. N.Y. Acad. Sci.* 1960; 85: 610-629.

[175] Smith, D; Ryan, M J; Day, JG. *The UKNCC Biological Resource: Properties, Maintenance, and Management*; Egham, Surrey, UK: UKNCC Secretariat; 2001.

[176] Smith, D; Thomas, VE. Cryogenic light microscopy and the development of cooling protocols for the cryopreservation of filamentous fungi. *World J.Microbiol. Biotechnol.* 1998; 14: 49-57.

[177] Meryman, HT; Williams, RJ; Douglas, MS. Freezing injury from "solution effects" and its prevention by natural or artificial cryoprotection. *Cryobiology* 1977; 14: 287- 302.

[178] Mazur, P. Survival of fungi after freezing and dessication. In: Ainsworth, GC; Sussman, AS (eds.) *The fungi - an advanced treatise,* vol. 3; New York - London: Academic Press; 1968; 325-394.

[179] Calcott, PH. *Freezing and thawing microbes*. Durham, UK: Meadowfield Press; 1978.

[180] Meryman, HT. Mechanics of freezing in living cells and tissues. *Science* 1956; 124: 515-521.

[181] Cunningham, JL. Preservation of rust fungi in liquid nitrogen. *Cryobiology* 1973; 10: 361-363.

[182] Fujikawa, S; Miura, K. Ultrastructural preservation of plasma membranes by non- lethal slow freezing to liquid nitrogen temperature. *Cell Struct. Funct.* 1987; 12: 63- 72.

[183] Wellman, AM; Pendyala, L. Permeability changes in membranes of *Neurospora crassa* after freezing and thawing. *Cryobiology* 1979; 16: 184-195.

[184] Wellman, AM. Growth of some fungi before and after cryogenic storage. In: Iizuka, H; Hasegawa, T (eds); *Culture collections of microorganisms*; Tokyo: Univ. Press; 1970; 255-264.

[185] Wellman, AM. Penicillin titer after storage of *Penicillium chrysogenum* by liquid nitrogen refrigeration. *Cryobiology* 1971; 7: 259-262.

[186] Barnhart, ER; Terry, CE. Cryobiology of *Neurospora crassa*. I. Freeze response of *Neurospora crassa* conidia. *Cryobiology* 1971; 8: 323-327.

[187] Schipper, MAA; Bekker-Holtman, J. Viability of lyophilized fungal cultures. *Antonie van Leewenhoek* 1976; 42: 325-328.

[188] Smith, D; Ward, SM. *Notes on the preservation of fungi*. Slough, UK: CAB International Mycological Institute; 1987.

[189] Ryan, MJ; Smith, D. Preserving a microorganism´s capacity to produce active biomolecules. In: Lima, N; Smith, D (eds); *Biological resource centres and the use of microbes*; Braga, Portugal: Livraria Minho; 2003; 197-216.

[190] Tan, CS; Stalpers JA. Vitrification of fungi; In: Samson RA; Stalpers, JA; van der Mei, D; Stouthamer, AH (eds); *Culture collections to improve the quality of life. Proceedings of the Eight International Congress for Culture Collections.* Veldhoven, The Netherlands, 25-29 August 1996. Baarn, The Netherlands: Centraalbureau voor Schimmelcultures and the World Federation for Culture Collections; 441.

[191] Tan, CS; Stalpers, JA. Vitrification of fungi. In: Cimerman, A; Gunde-Cimerman, N (eds); *Biodiversity, International Biodiversity Seminar ECCO XIV Meeting*, Ljubljana, Slovenia: ECCO; 1996; 189-193.

[192] Benson, EE; Wilkinson, M; Todd, A; Ekuere, U; Lyon, J. Developmental competence and ploidy stability in plants regenerated from cryopreserved potato shoot tips. *CryoLetters* 1996; 17: 119-128.

[193] Ryan, M J. The use of immobilisation for the preservation of *Serpula lacrymans. Mycologist* 2001; 15: 66-69.

[194] Declerck, S; Angelo-Van Coppenolle, MG. Cryopreservation of entrapped monoxenically produced spores of an arbuscular mycorrhizal fungus. *New phytologist* 2000; 148: 169-176.

[195] Nussinovitch, A. *Polymer macro- and micro-gel beads: fundamentals and applications*. New York: Springer; 2010; 231 pp.

[196] Coulson, GE; Morris, GJ; Smith, D. A cryomicroscopic study of *Penicillium expansum* hyphae during freezing and thawing. *J. Gen. Microbiol.* 1986; 132: 183- 190.

[197] Morris, GJ; Smith, D; Coulson, GE. A comparative study of the morphology of hyphae during freezing with the viability upon thawing of 20 species of fungi. *J. Gen. Microbiol.* 1988; 134: 2897-2906.

[198] Thomas, V; Smith, D. Cryogenic light-microscopy and the development of long-term cryopreservation techniques for fungi. *Outlook on Agriculture* 1994; 23: 163-167.

[199] Refshauge, S; Watt, M; McCully, ME; Huang, CX. Frozen in time: a new method using cryo-scanning electron microscopy to visualize root-fungal interactions. *New Phytologist* 2006; 172: 369-374.

[200] Smith, D; Ryan, MJ. Current status of fungal collections and their role in biotechnology. In: Arora, D K (ed.); *Handbook of Fungal Biotechnology, 2nd ed.*; New York: Marcel Dekker, Inc.; 2004; 527-538.

[201] Ryan, MJ; Ellison, CA. Development of a cryopreservation protocol for the microcyclic rust-fungus *Puccinia spegazzinii. CryoLetters* 2003; 24: 43-48.

[202] Horgen, P; Carvalho, D; Sonnenberg, AM; Li, A; van Griensven, LJLD. Chromosomal abnormalities associated with strain degeneration in the

cultivated mushroom, *Agaricus bisporus*. *Fung. Genet. Biol.* 1996; 20: 229-241.

[203] Kuhls, K; Liekfeldt, E; Borner, T. PCR fingerprinting used for comparison of ex type strains of *Trichoderma* species deposited in different culture collections; *Microb. Res.* 1995; 150: 363-371.

[204] Ryan, MJ; Bridge, PD; Smith, D; Jeffries, P. Phenotypic degeneration occurs during sector formation in *Metarhizium anisopliae*. *J. Appl. Microbiol.* 2002; 93, 163-168.

[205] Santos, IM; Abrunhosa, L; Venancio, A; Lima, N. The effect of culture preservation techniques on patulin and citrinin production by *Penicillium expansum* Link. *Lett. Appl. Microbiol.* 2002; 35: 272-275.

[206] Clutterbuck, AJ. Stability and instability of fungal genomes. In: Arora, DK (ed.). *Handbook on fungal biotechnology, 2nd edn.* New York: Marcel Dekker; 2004; 69-80.

[207] Lopez Lastra, CC; Hajek, AE; Humber, RA. Effects of two cryopreservation techniques on viability and pathogenicity of entomophthoralean fungi. *Can. J. Bot.* 2001; 79: 861-864.

[208] Horita, M; Kita, K; Tsuchiya, K. Stable preservation of pathogenicity of *Fusarium moniliforme* strains by freeze-drying method. *Proc. Kanto Tosan Plant Protect. Soc.* 1994; 41: 13-14.

[209] Ohmasa, M; Tsunoda, M; Babasaki, K; Hiraide, M; Harigae, H. Fruit-body production of test cultures of *Flammulina velutipes* preserved for seven years by freezing at three different temperatures. *Mycoscience* 1996; 37: 449-454.

[210] Suman, BC; Jandaik, GL. Preservation of cultures *Agaricus bisporus* (Lange) Sing. in liquid nitrogen and its effect on yield and characters of fruiting bodies. *Indian J. Mycol. Pl. Pathol.*1991; 21: 34-37.

[211] Jodon, MH; Royse, DJ; Jong, SC. Productivity of *Agaricus brunnescens* stock cultures following 5-, 7-, and 10-year storage periods in liquid nitrogen. *Cryobiology* 1982; 19: 602-606.

[212] Dupont, J; Joly, P; Roquebert, MF; Debievre, C. Preservation and instability of filamentous fungi strains - effect of cryopreservation and freeze-drying. *Bull. Soc. Bot. France-Actual. Bot.* 1986, 133: 105-124.

[213] Gaylarde, C; Kelley, J. Genetic variation in *Fusarium merismoides* preserved by two different methods. *World J. Microbiol. Biotechnol.* 1995; 11: 319-321.

[214] Butterfield, W; Jong, SC. Retention of albino and brown phenotypes of *Histoplasma capsulatum* by liquid nitrogren refrigeration. *Sabouraudia* 1975; 13: 291-294.

[215] Day, AW; Wellman AM; Martin, J. Recombination in *Ustilago violacea* after liquid nitrogen refrigeration. *Can. J. Microbiol.* 1972; 18: 1939-1641.

[216] MacDonald, KD. Storage of conidia of *Penicillium chrysogenum* in liquid nitrogen. *Appl. Microbiol.* 1972; 23: 990-993.

[217] Homolka, L. Production variability in *Oudemansiella mucida. Folia Microbiol.* 1977; 22: 484-485.

[218] Tan, CS; Van Ingen, CW. Preservation of fungi and yeast. In: Fuller, BJ; Lane, N; Benson, EE (eds); *Life in the Frozen State*; Boca Raton, FL: CRC Press; 2004; 277- 299.

[219] Cox, SL; Hulston, D; Maas, EW. Cryopreservation of marine thraustochytrids (*Labyrinthulomycetes*). *Cryobiology* 2009; 59: 363-365.

[220] Yarlett, NC; Yarlett, N; Orpin, CG; Lloyd, D. Cryopreservation of the anaerobic rumen fungus *Neocallimastix patriciarum. Lett. Appl. Microbiol.* 1986; 3: 1-3.

[221] Bromfield, KR; Schmitt, CG. Cryogenic storage of conidia of *Peronospora tabacina. Phytopathology* 1967; 57: 1133.

[222] Kilpatrick, RA; Harmon, DL; Loegering, WQ; Clark, WA. Viability of ureidospores of *Puccinia graminis* f.sp. *tritici* stored in liquid nitrogen. *Plant Dis. Rep. 1960-1970.* 1971; 55: 871-873.

[223] Cunningham, JL. Longevity of rust spores in liquid nitrogen. *Plant Dis. Rep.* 1973; 57: 793-795.

[224] Wyllie, TD; Fry, G. Liquid nitrogen storage of *Macrophomina phaseolina. Plant Dis. Rep.* 1973; 57: 478-480.

[225] Gale, AW; Schmitt, CG; Bromfield, KR. Cryogenic storage of conidia of *Sclerospora sorghi. Phytopathology* 1975; 65: 828-829.

[226] Sanchez-Pena, SR; Thorvilson, HG. Effect of long-term cryogenic storage and conidial suspending agents on the virulence of *Beauveria bassiana* toward *Solenopsis invicta. J. Invertebr. Pathol.* 1995; 65: 248-252.

[227] Georgopoulos, A. Tiefgefrierkonservierung von Pilzen in flüssigem Stickstoff als Grundlage für standardisierte Inokula. *Mykosen* 1978; 21: 19-23.

[228] Boyle, DG; Hyatt, AD; Daszak, P; Berger, L; Longcore, JE; Porter, D.; Hengstberger, SG; Olsen, V.: Cryo-archiving of *Batrachochytrium dendrobatidis* and other chytridiomycetes. *Dis. Aquat. Organisms* 2003; 56: 59-64.

[229] Tyrrell, D; Sohi, SS; Welton, MA. Preservation of *Entomophtora* protoplasts in liquid nitrogen. *Can. J. Microbiol.* 1972; 18: 1967-1968.

[230] Wood, CB; Pritchard, HW; Miller, AP. Simultaneous preservation of orchid seed and its fungal symbiont using encapsulation-dehydration is dependent on moisture content and storage temperature. *CryoLetters* 2000; 21: 125-136.

[231] Honegger, R. The impact of different long-term storage conditions on the viability of lichen-forming ascomycetes and their green algal photobiont, *Trebouxia* spp. *Plant Biol.* 2003; 5: 324-330.

[232] Ochi, S; Nakagawa, A. A simple method for long-term cryopreservation of *Calonectria ilicicola* on barley grains. *J. Gen. Plant Pathol.* 2010; 76: 112-115.

[233] Holden, ANG; Smith, D. Effects of cryopreservation methods in liquid-nitrogen on viability of *Puccinia abrupta* var. *partheniicola* urediniospores. *Mycol. Res.* 1992, 96: 473-476.

[234] Legard, DE; Chandler, CK. Cryopreservation of strawberry pathogens in a -95°C mechanical ultra-low temperature freezer. *Hort Sci.* 2000; 35: 1357.

[235] Zhu, FX; Bryson, PK; Schnabel, G. Influence of storage approaches on instability of propiconazole resistance in *Monilinia fructicola*. *Pest Manag. Sci.* 2012; 68: 1003- 1009.

[236] Webb, KM; Hill, AL; Laufman, J; Hanson, LE; Panella, L. Long-term preservation of a collection of *Rhizoctonia solani* using cryogenic storage. *Ann. Appl. Biol.* 2011; 158: 297-304.

[237] Perez-Garcia, A; Mingorance, E; Rivera, ME; Del Pino, D; Romero, D; Tores, JA; De Vicente, A. Long-term preservation of *Podosphaera fusca* using silica gel. *J. Phytopathol.* 2006; 154: 190-192.

[238] ten Hoopen, GM; Ortiz, JL; Aguilar, ME; Krauss, U. Preservation methodology for *Rosellinia* species. *Mycol. Res.* 2004; 108: 274-282.

[239] Widmer, TL. Cryopreservation of *Synchytrium solstitiale* in planta. Plant Dis. 2006; 90: 429-432.

[240] Shimizu, S; Hayata, C. Cryopreservation of protoplasts of an entomogenous fungus, *Paecilomyces fumosoroseus*. *J. Invertebr. Pathol.* 1990; 56: 283-285.

[241] Sandskar, B; Magalhaes, B. Cryopreservation of *Zoophthora radicans* (*Zygomycetes*, *Entomophthorales*) in liquid-nitrogen. *Cryobiology* 1994; 31: 206-213.

[242] Faria, MR; Martins, I; Mello, R; Tigano, MS. Entomopathogenic fungal (hyphomycetes) collection: Assessment of conidial viability. *Pesq. Agr. Brasil.* 1999; 34: 1497-1503.

[243] Oliveira, I; Pereira, JA; Bento, A; Baptista, P. Viability of *Beauveria bassiana* isolates after storage under several preservation methods. *Ann. Microbiol.* 2011, 61: 339-344.

[244] Humber, RA. Fungi: preservation of cultures; In: Lacey, LA (ed.); *Manual of techniques in insect pathology*. London: Academic Press; 1997; 269-280.

[245] Butt, TM; Hajek, AE; Humber, RA. The effect of temperature on growth and survival of protoplasts of the gypsy moth pathogen, *Entomophaga maimaiga*. *J. Invertebr. Pathol.* 1994; 64: 74-75.

[246] Dlugonski, J; Wilmanska, D. Application of protoplasts from *Curvularia lunata* for steroid hydroxylation and genetic transformation. *Microbios* 1998; 94: 113-121.

[247] Dlugonski, J; Slaba, M; Wilmanska, D. The use of cryopreserved apical protoplasts from *Curvularia lunata* for electrotransformation. *Microbios* 1998; 95: 71-77.

[248] Mota, MA; Campos, AK; Araujo, JV. Influence of different storage methods on the predatory capacity of the fungi *Arthrobotrys robusta* and *Monacrosporium thaumasium* after passage through the bovine gastrointestinal tract. *World J. Microbiol. Biotechnol.* 2003; 19: 913-916.

[249] Mota, MA; Campos, AK; Araujo, JV. Sporulation, radial growth and biomass production of *Arthrobotrys robusta* and *Monacrosporium thaumasium* submitted to different methods of preservation. *Brazil. J. Microbiol.* 2003; 34: 157-160.

[250] Baker, M; Jeffries, P. Use of commercially available cryogenic vials for long-term preservation of dermatophyte fungi. *J. Clin. Microbiol.* 2006; 44: 617-618.

[251] Espinel-Ingroff, A; Montero, D; Martin-Mazuelos, E. Long-term preservation of fungal isolates in commercially prepared cryogenic microbank vials. *J. Clin. Microbiol.* 2004; 42: 1257-1259.

[252] Hasegawa, A. The preservation of dermatophytes in parasitic form by freezing and freeze-drying. *Cryobiology* 1973; 10: 375-378.

[253] Delalibera, I; Humber, RA; Hajek, AE. Preservation of *in vitro* cultures of the mite pathogenic fungus *Neozygites tanajoae*. *Can. J. Microbiol.* 2004; 50: 579-586.

[254] Jensen, AB; James, RR; Eilenberg, J. Long-term storage of *Ascosphaera aggregata* and *Ascosphaera apis*, pathogens of the leafcutting bee (*Megachile rotundata*) and the honey bee (*Apis mellifera*). *J. Invertebr. Pathol.* 2009; 101: 157-160.

[255] Gleason, FH; Mozley-Standridge, SE; Porter, D; Boyle, DG; Hyatt, AD. Preservation of *Chytridiomycota* in culture collections. *Mycol. Res.* 2007; 111: 129-136.

[256] Missous, G; Thammavongs, B; Dieuleveux, V; Gueguen, M; Panoff, JM. Improvement of the cryopreservation of the fungal starter *Geotrichum candidum* by artificial nucleation and temperature downshift control. *Cryobiology* 2007; 55: 66-71.

[257] Missous, G; Thammavongs, B; Launaya, G; Guéguen, M; Dieuleveux, V; Panoff, JM. Relationship between growth behaviour, micro and macroscopic morphologies and freezing sensitivity of the ripening starter *Geotrichum candidum* is strain specific and mostly related to the morphotypes: the arthrospores/hyphae parameter. *J. Dairy Res.* 2010; 77: 425-431.

[258] Antheunisse, J. Viability of lyophilized microorganisms after storage. *Antonie van Leeuwenhoek* 1973; 39: 243-248.

[259] Hwang, SW; Kwolek, WF; Haynes, WC. Investigation of ultra low temperature for fungal cultures. III. Viability and growth rate of mycelial cultures following cryogenic storage. *Mycologia* 1976; 68: 377-387.

[260] Bazzigher, G. Ein vereinfachtes Gefriertrocknungsverfahren zur Konservierung von Pilzkulturen. *Phytopathol. Z.* 1962; 45: 63-56.

[261] Tommerup, IC. Long-term preservation by L-drying and storage of vesicular arbuscular mycorrhizal fungi. *Trans. Brit. Mycol. Soc.* 1988; 90: 585-591.

[262] Danell, E; Flygh, G. Cryopreservation of the ectomycorrhizal mushroom *Cantharellus cibarius*. *Mycol. Res.* 2002; 106: 1340-1342.

[263] Chvostová, V; Nerud, F; Homolka, L. Viability of wood-inhabiting basidiomycetes following cryogenic preservation. *Folia Microbiol.* 1995; 40: 193-197.

[264] Ito, T. Preservation of fungal cultures at the Institute of Fermentation, Osaka (IFO), In: Samson, RA; Stalpers, JA; van der Mei, D; Stouthamer, AH (Eds); *Culture collections to improve the quality of life (Proceedings of the Eight International Congress for Culture Collections,* Veldhoven, The Netherlands, 25-29 August 1996); Baarn, The Netherlands: Centraalbureau voor Schimmelcultures and the World Federation for Culture Collections; 1996; 210-211.

[265] San Antonio, JP; Hwang, SC. Liquid nitrogen preservation of *Agaricus bisporus* (Lange) Sing. spawn stocks. *Mushroom J.* 1982; 120: 410-419.

[266] Maekawa, N; Fukuda, M; Arita, I; Komatsu, M. Cryopreservation of edible basidiomycetous fungi in liquid nitrogen. *Rep. Tottori Mycol. Inst.* 1988; 26: 15-28.

[267] Maekawa, N; Fukuda, M; Arita, T; Komatsu, M. Effects of liquid-nitrogen cryopreservation on stock cultures of three cultivated basidiomycetous fungi. *Rep. Tottori Mycol. Inst.* 1990; 28: 227-232.

[268] Lara-Herrera, I; Mata, G; Gaitán-Hernández, R. Evaluation of the viability of *Pleurotus* spp.strains after liquid nitrogen cryopreservation. *Rev. Microbiol.* 1998; 29: 193-196.

[269] Mata, G; Salmones, D; Ortega, PM. Viability and mushroom production of *Lentinula edodes* and *L. boryana* strains (Fungi: basidiomycetes) after cryogenic storage of spawn stocks. *World J. Microbiol. Biotechnol.* 2000; 16: 283-287.

[270] Kitamoto, Y; Suzuki, A; Shimada, S; Yamanaka, K. A new method for the preservation of fungus stock cultures by deep-freezing. *Mycoscience* 2002; 43: 143- 149.

[271] Karaduman, AB; Atli, B; Yamac, M. An example for comparison of storage methods of macrofungus cultures: *Schizophyllum commune*. *Turk. J. Bot.* 2012; 36: 205-212.

[272] Colauto, NB; Cordeiro, FA; Geromini, KVN; de Lima, TG; Lopes, AD; Nunes, RAR; Roratto, FD; Tanaka, HS; Zaghi, LL; Linde, GA. Viability of *Agaricus blazei* after long-term cryopreservation. *Ann. Microbiol.* 2012; 62: 871-876.

[273] Colauto, NB; Eira, AF; Linde, GA. Cryopreservation at -80°C of *Agaricus blazei* on rice grains. *World J. Microbiol. Biotechnol.* 2011; 27: 3015–3018.

[274] Homolka, L; Lisá, L; Nerud, F. Basidiomycete cultures on perlite survive successfully repeated freezing and thawing in cryovials without subculturing. *J. Microbiol. Meth.* 2007; 69: 529-532.

[275] Sato, M; Sukenobe, J; Nakagiri, A. Cryopreservation of cryosensitive basidiomycete cultures by application and modification of perlite protocol. *CryoLetters* 2012; 33: 87- 95.

[276] Hutchison LJ. Absence of conidia as a morphological character in ectomycorrhizal fungi. *Mycologia* 1989; 81: 587-594.

[277] Heinonen-Tanski, H. Maintenance methods for ectomycorrhizal fungi. *Agriculture, Ecosystems and Environment* 1990; 28: 171-174.

[278] Molina, R; Caldwell, BA; Castellano, MA; Horton, TR; Smith, JE. Mycorrhizae: ectomycorrhizal fungi. In: Bitton, G (ed.); *Encyclopedia of Environmental Microbiology*. New York: Wiley; 2002; 2124-2132.

[279] Dalpé, Y; Monreal, M. Arbuscular mycorrhiza inoculum to support sustainable cropping systems. Online. *Crop Management* doi:10.1094/CM-2004-0301-09-RV; Plant Management Network; 2004.

[280] Kuszala, C; Gianinazzi, S; Gianinazzi-Pearson, V. Storage conditions for the long- term survival of AM fungal propagules in wet sieved soil fractions. *Symbiosis* 2001; 30: 287-299.

[281] [281 Sommerville, KD; Siemon, JP; Wood, CB; Offord, CA. Simultaneous encapsulation of seed and mycorrhizal fungi for long-term storage and propagation of terrestrial orchids. *Austr. J. Bot.* 2008; 56: 609-615.

[282] Batty, AL; Dixon, KW; Brundrett, M; Sivasithamparam, K. Long-term storage of mycorrhizal fungi and seed as a tool for the conservation of endangered Western Australian terrestrial orchids. *Austr. J. Bot.* 2001; 49: 619-628.

[283] Stielow, JB; Vaas, LA; Goeker, M; Hoffmann, P; Klenk, HP. Charcoal filter paper improves the viability of cryopreserved filamentous ectomycorrhizal and saprotrophic *Basidiomycota* and *Ascomycota*. *Mycologia* 2012; 104: 324-330.

[284] Obase, K; Lee, SY; Chun, KW; Lee, JK. Regeneration of ectomycorrhizal fungal isolates following deep freezer storage. *Mycobiology* 2011; 39: 133-136.

[285] Thomson, BD; Malajczuk, N; Grove, TS; Hardy, GEStJ. Improving the colonization capacity and effectiveness of ectomycorrhizal fungal cultures by association with a host plant and re-isolation. *Mycol. Res.* 1993; 97: 839-844.

[286] Mata, G; Perez-Merlo, R. Spawn viability in edible mushrooms after freezing in liquid nitrogen without a cryoprotectant. *Cryobiology* 2003; 47: 14-20.

[287] Hwang, SW; San Antonio, JP. Stability of spawn stock of the cultivated mushroom after 26 months liquid nitrogen refrigeration (-160°C to -196°C). *Mushroom Sci.* 1972; 8: 35-42.

[288] Kochkina, GA; Ivanushkina, NE; Karasev, SG; Gavrish, EY; Gurina, LV; Evtushenko, LI; Spirina, EV; Vorob'eva, EA; Gilichinskii, DA; Ozerskaya, SM. Survival of micromycetes and actinobacteria under conditions of long-term natural cryopreservation. *Microbiology* 2001; 70: 356-364.

In: Cryopreservation
Editors: A. Colvert and H. Coty

ISBN: 978-1-62618-474-9

Chapter 2

DIFFERENTIAL SCANNING CALORIMETRY: A USEFUL TOOL FOR BETTER UNDERSTANDING CRYSTALLISATION AND SLOW-FREEZING PROTOCOLS

P. Bruyere[1], A. Baudot[2], P. Guérin[1], G. Louis[2], T. Joly[1] and S. Buff[1]

[1]VetAgroSup – Campus Vétérinaire de Lyon, Université de Lyon, Marcy l'Etoile, France

[2]Université Paris-Descartes, Sorbonne Paris Cité, Paris, France

Abstract

The control of ice formation and suitable cooling rates are fundamental conditions for the success of slow-freezing protocols. However, only biological studies allow usually evaluation of such protocols, which makes the results dependent on factors external to the congelation.

Consequently, fundamental analytical methods, especially differential scanning calorimetry, have improved our understanding of cryopreservation protocols. Differential scanning calorimetry allows thermodynamic evaluation of cryopreservation solutions under dynamic conditions. Consequently, thermodynamic characteristics, such as the crystallisation temperature, amount of ice crystallised and melting temperature, can be evaluated. In this chapter, after explaining the principle of a differential scanning calorimeter, we will review the main differential scanning calorimetry studies to better understand slow-freezing protocols and crystallisation. In particular, we will focus on the characterisation of slow-freezing solutions (part II), optimisation of slow-freezing protocols (part III), elaboration of phase diagrams of solutions containing cryoprotective agents (part IV) and characterisation of intracellular crystallisation (part V).

Abbreviations

BSA:	bovine serum albumin
CPAs:	cryoprotective agents
DMSO:	dimethyl sulfoxide
DSC:	differential scanning Calorimetry
EG:	ethylene glycol
FCS:	fœtal calf serum
ΔH:	crystallisation enthalpy
Q:	heat of ice crystallisation
T_c:	crystallisation temperature
T_m:	end of melting temperature
T_{ph}:	phase transition temperature

Introduction

Slow-freezing protocols are characterised by control of ice formation. Currently, these protocols are mainly evaluated through biological studies in which their efficiencies are determined with regard to cellular viability. However, many external factors can have an impact on cellular viability and thus alter the conclusions drawn from these studies. Consequently, the understanding of slow-freezing protocols can be improved by more fundamental methods that allow elimination of many external factors.

Differential scanning calorimetry is a fundamental analytical method that has already been used in the evaluation of crystallisation during slow-freezing protocols. In fact, because crystallisation and melting are exothermic and endothermic phenomena, they can be evaluated by thermodynamic analyses. In this chapter, we will review the main studies in which differential scanning calorimetry (DSC), a calorimetric analytical method, has been used to better understand crystallisation and slow-freezing protocols. Two categories of DSC exist: power compensation DSC and heat flux DSC. Because most cryobiology studies use power compensation DSC, we will limit the review to this analytical method.

The Principle of Power Compensation Differential Scanning Calorimetry

In DSC studies, phase transitions are characterised by measuring heat exchanges between the sample and an adiabatic chamber, called the "measuring head". These measurements are performed during cooling and heating at a constant rate.

The measuring head of a power compensation DSC is composed of two microfurnaces placed in a chamber that is continuously cooled (Figure 1). This cooling is provided by an aluminium radiator that is placed under the microfurnaces and plunged in liquid nitrogen. Homogenisation of the microfurnace cooling is realised by a helium gas flux in the chamber. A temperature sensor and a heating resistor are placed under each microfurnace and allow the precise measure and precise adjustment, respectively, of the temperature of each microfurnace independently of the other.

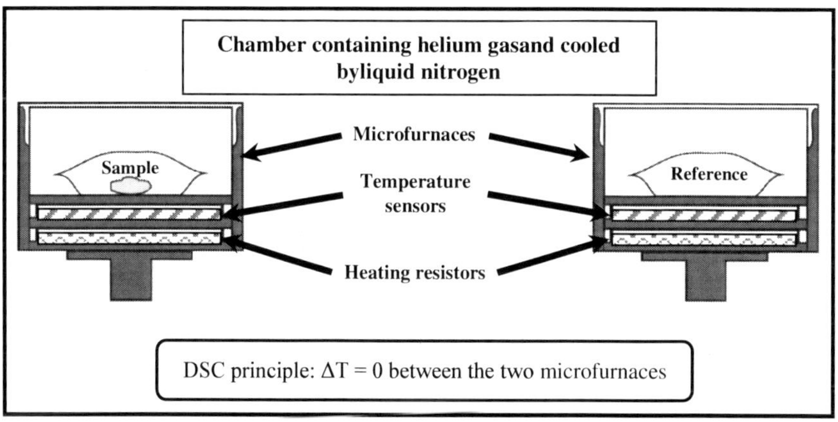

Figure 1.The DSC principle. *The principle of power compensation DSC is to maintain the same temperature between the two microfurnaces whatever the phase transitions occurring in the sample. When a phase transition occurs, the heats provided by the two heating resistors are different. This difference allows characterisation of the phase transition.*

One microfurnace contains the sample, and the other contains a reference. The principle of power compensation DSC is to maintain the same temperature between the two microfurnaces whatever the phase transitions occurring in the sample (Clas et al., 1999). When a phase transition occurs in the sample, the heat that is absorbed or released by the sample is compensated by adding or removing an equivalent amount of heat in the heating resistor, which is placed under the microfurnace containing the sample. Consequently, the powers provided by the two heating resistors are different. Thus, in a power compensation DSC, the measured variable is the power difference supplied by the two resistors to maintain the same temperature between the two microfurnaces (Jiang et al., 2002). This difference is materialised on a curve called a thermogram and allows characterisation of the phase transition (Figure 2).

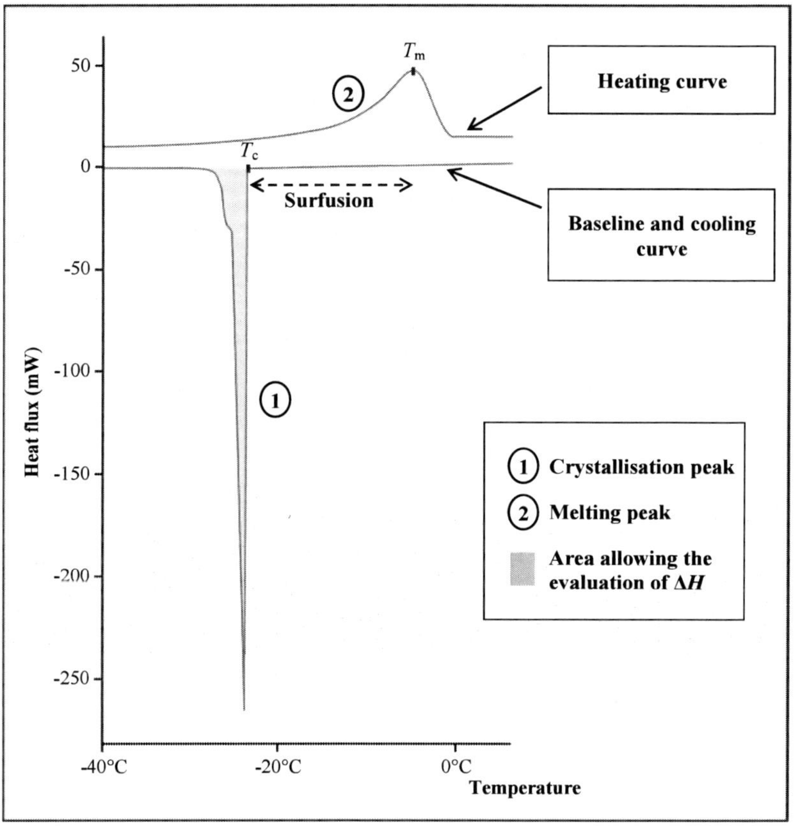

Figure 2.Thermogram and analysis scheme. *Several calorimetric characteristics can be measured on a thermogram:*

- T_c*: the crystallisation temperature.*
- ΔH*: the crystallisation enthalpy, which is proportional to the amount of ice crystallised in the solution.*
- T_m*: the end of melting temperature.*

On a thermogram, the power difference is expressed as a function of temperature in mW. When no phase transition occurs in the sample, the powers of the two heating resistors are almost identical. Consequently, the curve has the shape of an almost horizontal line called the baseline. When a phase transition occurs, the heat of the resistor under the sample is modified, and a difference between the heats of the two resistors appears. A phase transition is thus indicated by a peak on the thermogram. When the phase transition is exothermic (crystallisation), the heat provided by the resistor under the sample decreases. The difference between the two heats is negative, and the peak is directed downward. When the phase transition is endothermic (ice melting), the heat provided by the resistor under the sample increases, and the peak is directed upward.

Finally, phase transitions are not isothermal on the thermogram and are represented by flared shape peaks because of the dynamic aspect of the measurement. Indeed, the measurements are conducted during cooling and warming at a constant rate, which is imposed by the DSC. However, the thermal diffusivity of the sample prevents an instantaneous phase transition of the whole sample. Consequently, the phase transition cannot be isothermal. Because the crystallisation phenomenon is more abrupt than the ice melting phenomenon, the crystallisation peak is sharper than the melting peak.

Several thermodynamic characteristics can be measured on a thermogram. In particular, the crystallisation temperature (T_c) and crystallisation enthalpy (ΔH) can be evaluated during slow cooling. T_c is measured at the beginning of the crystallisation peak, whereas ΔH is measured using the area encompassed between the crystallisation peak and baseline. During a slow warming, the end of melting temperature (T_m) can be measured at the top of the melting peak.

To conclude, ΔH can be used to calculate the maximal heat of ice crystallisation (Q_{max}) as follow:

$$\boxed{Q_{max}(\%) = \frac{\Delta H}{\Delta H_{water}} \times 100}$$

with $\Delta H_{water} = 333.0\ J.g^{-1}$. Q_{max} evaluates the mass (g) of ice whose solidification at 0°C would liberate the same amount of heat as a mass of 100 g of solution on crossing the corresponding peak (Baudot et al., 2000). Thus, Q_{max} is close to the maximal quantity of ice crystallised by the percentage of solution ($w.w^{-1}$) (Boutron, 1984).

Characterisation of Slow-Freezing Solutions

Only two studies have focused on the characterisation of slow-freezing solutions with DSC. However, a third study in which saline solutions containing 10% glycerol have been characterised will also be described.

In the first study, three cryopreservation solutions for rabbit sperm were evaluated to determine the influence of antibiotic additions on their thermodynamic properties (Salvetti et al., 2006). T_c and Q_{max} were used for the comparisons. Salvetti et al. first observed that the calorimetric properties of the solutions without antibiotics were similar. However, the addition of antibiotics had various effects on the solutions. In two solutions, the addition of antibiotics promoted ice formation (increase of T_c and Q_{max}). However, this addition limited ice formation in the third solution.

The analysis of these results was difficult because of differences between the compositions of the solutions. Nevertheless, the authors proposed that the antibiotics could promote heterogeneous nucleation and thus facilitate crystallisation. In the first two solutions, the antibiotics could have interacted with the hydrophilic groups of the cryoprotectants. These interactions could have resulted in more “free water”, which could potentially crystallise. The third solution differed from the other two solutions through the presence of glycerol. It is possible that the glycerol, unlike other cryoprotective agents (CPAs), preferentially interacts with water and not with antibiotics. The amount of water potentially available to crystallise could thus be smaller than in the other solutions, which would explain the difference between their thermodynamic behaviours.

Therefore, the interaction between antibiotics and CPA was also evaluated in this study. Simplified solutions containing only distilled water and CPA corresponding to the slow-freezing solutions were created. The results showed that the decrease of crystallisation observed in the third solution was a consequence of the presence of glycerol associated with dimethyl sulfoxide(DMSO) and not due to the presence of glycerol alone. In these simplified solutions, the antibiotics tended to decrease the crystallisation, in contrast to the results obtained with two previously evaluated slow-freezing solutions. Consequently, the interactions between the various products of a cryopreservation solution are complex and have a significant role in the thermodynamic behaviour of the solutions.

In the second study, DSC was used to evaluate a chemically defined substitute for products of animal origin in bovine embryo cryopreservation solution (Bruyere et al., 2012). Slow-freezing solutions containing ethylene glycol

(EG) and saccharose associated with 20% fœtal calf serum (FCS), 20% of the potential substitute or 4 g.L^{-1} of bovine serum albumin (BSA) were first compared by DSC to evaluate if the thermodynamic properties of the substitute were similar to those of products of animal origin. Then, the solutions were compared using a classical biological approach (freezing/thawing of bovine embryos followed by *in vitro* culture).

In the calorimetric approach, T_m and ΔH were used as thermodynamic characteristics. The solutions containing FCS or the substitute had similar thermodynamic properties (mean and variability). The solution containing BSA had calorimetric properties that were similar for T_m and only slightly different for ΔH. In conclusion, the potential substitute can replace FCS or BSA in bovine embryo cryopreservation solutions from a thermodynamic point of view.

The biological approach was conducted using *in vitro*-produced bovine embryos. These embryos were frozen, stored in liquid nitrogen and then thawed before being cultured. After *in vitro* culture, the embryo survival rates obtained with the solutions containing FCS or BSA were similar. These results validated their use for reference cryoprotective solutions. Nevertheless, the embryo survival rate obtained with the solution containing the substitute was higher. Thus, the use of the substitute improved the survival of *in vitro*-produced bovine embryos after slow-freezing.

Consequently, the solution containing the substitute had thermodynamic properties similar to those of the solutions containing FCS or BSA and better biological properties. The substitute can consequently be used in *in vitro*-produced bovine embryo slow-freezing solutions to replace products of animal origin. Moreover, DSC can help researchers choose such substitutes by screening various potential products and selecting those that have thermodynamic properties similar to products of animal origin.

In a third study, DSC was used to evaluate, as a function of the cooling rate, the amount of crystallisation in a saline solution containing 10% glycerol (Morris et al., 2006). In the first part of the study, an analysis of the viscosity of the solution allowed the researchers to highlight a decrease in the mobility of water molecules between -7°C and -45°C. These results were not obtained with DSC. The mobility decrease was even more important than the cooling rate was high. After nucleation, ice crystal grow this achieved through the diffusion of water molecules from the solution to the crystal. A decrease in diffusion should consequently cause a decrease in ice crystal growth and thus in the amount of ice. This hypothesis was confirmed using DSC. When the cooling rate ranged between -1 and -5°C.min^{-1}, the amount of ice in the solution was 81% (w.w^{-1}). The amount of ice progressively decreased when the cooling rate increased and reached 55%

at -300°C.min^{-1}. From the saline solution results, the authors deduced that, during a fast cooling, both the decrease of the duration of dehydration and increase of the solution viscosity cause a decrease in the diffusion of water from the cell. These results reinforce the need to use a slow cooling rate during a slow-freezing protocol to allow sufficient dehydration of the cells.

Otherwise, during the warming, the amount of melted ice was consistently 81% for any cooling rate. The authors concluded that the possibility of ice crystal growth was insufficient during rapid cooling. Consequently, a recrystallisation phenomenon occurred during warming as soon as the mobility of the water molecules allowed it.

Optimisation of Slow-freezing Protocols

Several studies conducted by Devireddy have focused on optimising slow-freezing protocols using DSC experiments. During a slow-freezing process, control of the cooling rate after seeding allows intracellular water to diffuse progressively through the plasma membrane and crystallise in the extracellular medium. When cooling happens too fast, water cannot diffuse sufficiently to avoid intracellular crystallisation, and the cell dies.

In these studies, the optimal cooling rate was defined as the maximal cooling rate at which no intracellular ice crystallises (Devireddy et al., 1999). For one type of cell, this optimal rate depends on two criteria: the membrane permeability and the probability that the intracellular water remaining in the cell crystallises. The optimisation of the cooling rate and, consequently, of the slow-freezing protocol was performed by measuring the membrane permeability using DSC during a freezing process (Devireddy et al., 1998).

Crystallisation in cellular suspensions was evaluated using DSC (Devireddy et al., 1999). During the evaluation of one cellular suspension, two heats released by the suspension were measured successively:*(i)* the heat released during the freezing of osmotically active cells (living spermatozoa in which intracellular water diffuses through the plasma membrane and crystallises in the extracellular medium) and *(ii)* the heat released during the freezing of osmotically inactive cells (dead spermatozoa in which the intracellular water is mixed with the extracellular medium because of plasma membrane lysis). The difference between the two heats released was then correlated with the membrane permeability of osmotically active cells.

The two heats were evaluated using seven successive coolings and warmings:

1. A slow cooling (generally -5°C.min^{-1}) was first applied until the onset of extracellular crystallisation that was induced by extracts of *Pseudomonas syringae*. Indeed, this bacteria possesses nucleating agents that can trigger a relatively reproducible nucleation process (Kleinhans et al., 2006).
2. A slow warming was applied to reach the phase transition temperature (T_{ph}), which was not overshot in order to maintain some extracellular ice. Then, an isotherm at T_{ph} allowed preservation of the ice nuclei, but limited ice crystal growth and, consequently, osmotic shrinkage of cells.
3. A slow cooling (-5°C.min^{-1} or -20°C.min^{-1}) to -50°C allowed determination of the heat released by the extracellular medium containing the osmotically active cells ($Q_{initial}$).
4. A warming was applied until T_{ph}.
5. A rapid cooling (-200°C.min^{-1}) to temperatures below -100°C induced lysis of the cells, which became osmotically inactive.
6. A warming was applied until T_{ph}.
7. A slow cooling (-5°C.min^{-1} or -20°C.min^{-1}) to -50°C allowed the determination of the heat released by the extracellular medium containing the osmotically inactive cells (Q_{final}).

The total difference between the heats released during coolings three and seven, ΔQ (Figure 3), was calculated as follow:

$$\boxed{\Delta Q\ (\text{mW.mg}^{-1}) = Q_{initiale} - Q_{finale}}$$

A fractional difference between the heats released from T_{ph} until a temperature T, $\Delta Q(T)$ (Figure 4), was also calculated for different temperatures T.

ΔQ and $\Delta Q(T)$ were then correlated to the volume of cells at a temperature T, $V(T)$:

$$\boxed{V(T) = V_0 - \frac{\Delta Q(T)}{\Delta Q} \times \left(V_0 - V_b\right)}$$

with V_0 representing the initial or isotonic volume of cells and V_b the osmotically inactive volume of cells. The values of V_0 and V_b were obtained using the published values in the literature (Thirumala et al., 2003). The volumetric shrinkage of cells was then graphed and modelled using a nonlinear least squares curve-fitting method (Figure 5).

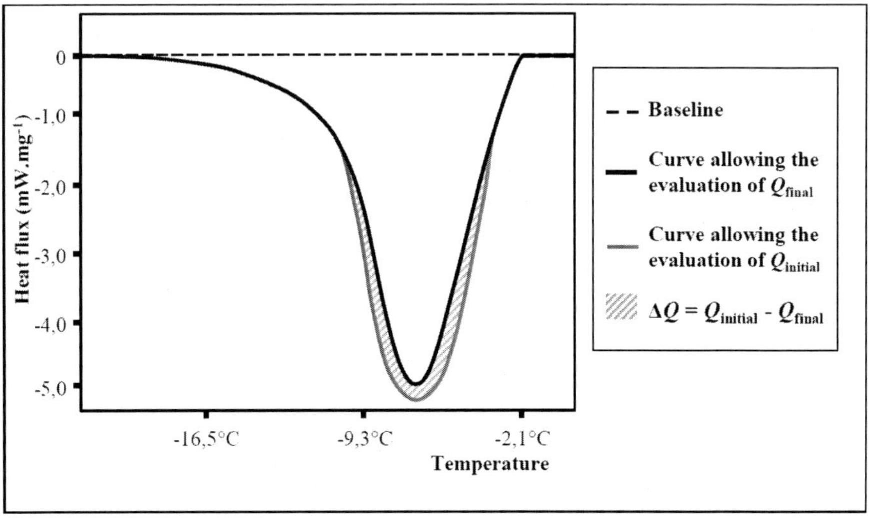

Figure 3.Scheme showing the thermograms allowing evaluation of Q_{initial} and Q_{final}. The superposition of the two thermograms allows calculation of the total difference between the heats released, ΔQ (mW.mg^{-1}; in strips), according to (Devireddy et al., 2004). *$Q_{initial}$: heat released by the cellular suspension when the cells are alive and thus osmotically active. Q_{final}: heat released by the cellular suspension when the cells are dead and thus osmotically inactive.*

The equation of the curve was deduced from the water transport model of Mazur, which was modified to incorporate the presence of CPA (Karlsson et al., 1994). This equation allowed the calculation of the cellular membrane permeability to water, L_{p}:

$$\frac{dV(T)}{dT} = -\frac{L_p A_c RT}{Bv_w} \times \left[\ln \frac{\left(V_0 - V_b - n_{\text{cpa}} v_{\text{cpa}}\right)/v_w}{\left(V_0 - V_b - n_{\text{cpa}} v_{\text{cpa}}\right)/v_w + \left(\varphi_s n_s + n_{\text{cpa}}\right)} - \frac{\Delta H_f v_w \rho}{R} \times \left(\frac{1}{T_R} - \frac{1}{T}\right) \right]$$

with A_c representing the effective membrane surface area for water transport, R representing the universal gas constant, B representing the constant cooling rate, v_w representing the molar volume of water, n_{cpa} representing the number of moles of CPA, v_{cpa} representing the molar volume of CPA, φ_s representing the disassociation constant of salt, n_s representing the number of osmoles of salt, ΔH_f representing the latent heat of fusion of water, ρ representing the density of water, and T_R representing the reference temperature (273.15 K).

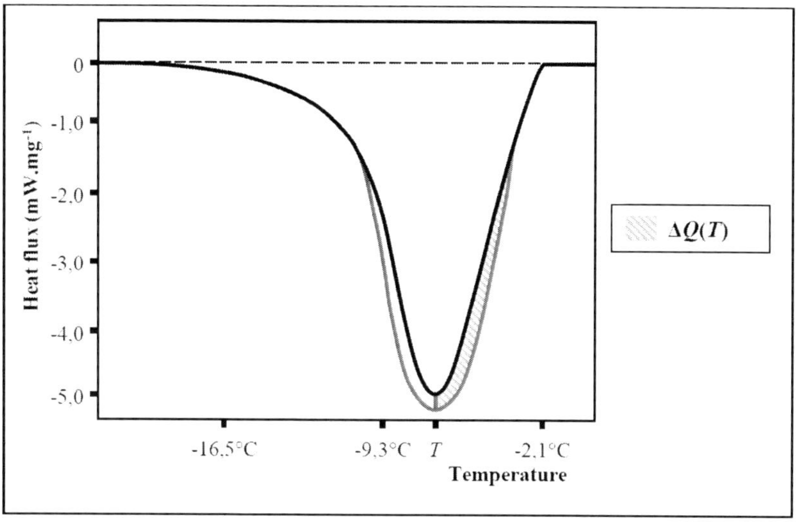

Figure 4.Scheme showing the thermograms allowing evaluation of Q_{initial} and Q_{final}. The superposition of the two thermograms allows calculation of the fractional difference between the heats released at temperature T, $\Delta Q(T)$(mW.mg^{-1}; in strips), according to (Devireddy et al., 2004). *ΔQ(T)(with stripes) corresponds to the difference of heats released between the phase transition temperature (T_{ph}) and a temperature T.*

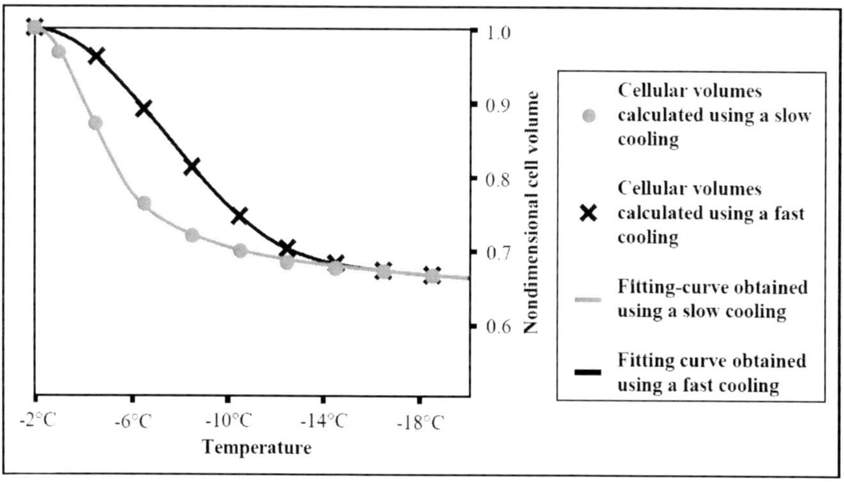

Figure 5. Scheme showing the fitting-curve of the cellular volumetric shrinkage depending on the cooling rate, according to (Devireddy et al., 2004). *For several temperatures, the cellular volume can be calculated using ΔQ and the ΔQ(T) obtained for these temperatures. The equation of the fitting-curve that models the volumetric shrinkage allows calculation of the plasma membrane permeability to water.*

L_p was then used to calculate the reference membrane permeability at 273.15 K, L_{pg}, and its apparent activation energy, E_{lp}:

$$L_p = L_{pg}\ exp\left[-\frac{E_{lp}}{R}\times\left(\frac{1}{T}-\frac{1}{T_R}\right)\right]$$

The replacement of L_{pg} and E_{lp} in the equation of the water transport model of Mazur was used to simulate the diffusion of intracellular water through the plasma membrane depending on the cooling rates. The authors could thus deduce an optimal cooling rate depending on the CPA concentrations.

This method allowed the evaluation of the optimal cooling rates for the freezing of different types of cells and tissues.

Table 1. Summary of the optimal cooling rates deduced from DSC measurements for mammal spermatozoa

Mammals	Mouse	Human	Horse	Dog	Boar	Bovine
Optimal cooling rate	26-44°C.min^{-1}	14-16°C.min^{-1}	29-60°C.min^{-1}	10-30°C.min^{-1}	30°C.min^{-1}	50-60°C.min^{-1}

Table 2. Summary of the optimal cooling rates deduced from DSC measurements for fish spermatozoa

Fish	Striped bass	Green swordtail	Southern platyfish	White bass
Optimal cooling rate	14-20°C.min^{-1}	20-35°C.min^{-1}	47°C.min^{-1} *	22°C.min^{-1}

* Discrepancy with experimental results.

Concerning cells, several studies focused on the cryopreservation of different species spermatozoa (Table 1 and Table 2): mouse (Devireddy et al., 1999), human (Devireddy et al., 2000), horse (Devireddy et al., 2002), dog (Thirumala et al., 2003), boar (Devireddy et al., 2004) and fish (Pinisetty et al., 2005; Thirumala et al., 2005; Devireddy et al., 2006; Thirumala et al., 2006). In most studies, the optimal cooling rates were in agreement with cooling rates that were previously published and empirically determined. A discrepancy was observed in only one study (Pinisetty et al., 2005), and this result could be due to a bad approximation in the intracellular crystallisation probability (calculation of the maximal volume of intracellular water remaining in the cell without causing intracellular crystallisation).

This method also allowed the comparison of the optimal cooling rates between ejaculated and epididymal spermatozoa. The authors observed that the optimal rates were similar in bovine (Alapati et al., 2009) and Rhesus monkey (Alapati et al., 2008) for both types of spermatozoa. Consequently, similar slow-freezing protocols could be used.

Finally, the optimal cooling rates were evaluated for the freezing of adipose derived adult stem cells (Thirumala et al., 2005). These rates ranged between -23.3°C.min^{-1} and -33.5°C.min^{-1} when CPAs were used and -22.1°C.min^{-1} and -96.7°C.min^{-1} when no CPA was used.

Regarding tissues, this method allowed the evaluation of the optimal cooling rates for the freezing of ovarian tissue. The optimal rates after induction of crystallisation are dependent on the cooling rates applied above 4°C for ovarian tissue of Rhesus monkeys (Li et al., 2006) and horses (Devireddy et al., 2006).

In a different field, uterine fibroids can be treated by cryosurgery. In such treatments, the freezing rate has to be high enough to induce lethal injuries to living tumorous cells. The authors observed that fibroid tumor tissue cells experienced essentially cellular dehydration when cooled at rates below -50°C.min^{-1} (Devireddy et al., 2001). Consequently, the cooling rates during cryosurgery should be higher than -50°C.min^{-1}.

To conclude, this method has been recently improved to be applied when intracellular ice forms (Mori et al., 2012). This improvement was based on the deconvolution of the curve allowing the evaluation of Q_{initial} into the heat releases due to:*(i)* the extracellular ice formation (Q_{EIF}), *(ii)* the water transport during the cell dehydration (Q_{WT}) and *(iii)* the intracellular ice formation (Q_{IIF}). On a practical point of view, Q_{EIF} was first calculated using an equation and was subtracted from Q_{initial}. Then, the curve obtained from this subtraction was deconvolved into two curves allowing the evaluation of Q_{WT} and Q_{IIF} (Figure 6).

The water transport evaluated with this new thermal analysis method was not different from that obtained with cryomicroscopy except for a cooling rate of -10°C.min^{-1}. One explanation for this discrepancy could be the simultaneous occurrence at this cooling rate of water transport and intracellular ice formation in a large fraction of cells. This phenomenon could lead to errors in the deconvolution of the thermogram. Further analysis of the deconvolved thermograms could improve the correlation between the water transports evaluated by DSC and cryomicroscopy. As intracellular ice formation was evaluated using this new analytical method, this study will also be discussed in the fifth part of this chapter.

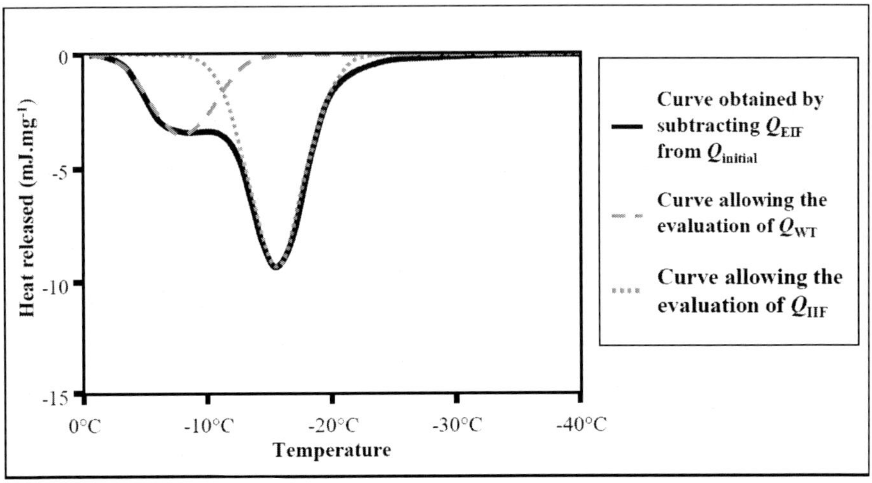

Figure 6. Scheme showing the curves allowing evaluation of Q_{WT} and Q_{IIF}, according to (Mori et al., 2012). $Q_{initial}$: heat released by the cellular suspension when the cells are alive and thus osmotically active. Q_{EIF}: heat released by the extracellular crystallisation. Q_{WT}: heat released by water transport. Q_{IIF}: heat released by the intracellular crystallisation.

Elaboration of Phase Diagrams of Solutions Containing Cryoprotective Agents

In cryobiology, phase diagrams of solutions containing CPA are very important and useful (Kleinhans et Mazur, 2007). These diagrams represent the different values of T_m as a function of the total weight percent concentration of solutes (CPA and salt). They depict, for any given negative temperature, the percentage of crystallised solution in equilibrium and, consequently, the concentration of solutes in the part of the solution that is still liquid. This concentration can then be used to calculate the osmotic pressure of the solution, which is one of the most essential factors for evaluating the dehydration of frozen cells. Currently, DSC is the most widely used method for determining the temperatures that are necessary to elaborate phase diagrams (Kleinhans and Mazur, 2007). Consequently, this analytical method is very helpful for examining slow-freezing protocols.

The most useful phase diagrams in cryobiology are those consisting of one or several CPAs dissolved in a physiological salt solution (Kleinhans et Mazur, 2007; Han et al., 2010; Weng et al., 2011). These diagrams represent the values of T_m as a function of the total weight percent concentration of solutes, and, for a given ratio $\frac{CPA\ weight}{NaCl\ weight}$. Consequently, determining these diagrams is relatively expensive and time-consuming in so far as T_m has to be defined for many concentrations of salts and CPAs. A phase diagram of a ternary system containing water, NaCl and a CPA (ternary phase diagram) requires the elaboration of several hundred samples (Han et al., 2010). The problem is even more important for saline solutions containing two CPAs (quaternary phase diagrams). New mathematical methods have been recently developed to determine complex phase diagrams from binary phase diagrams (water + CPA or water + NaCl), which are simpler to elaborate. These methods use complex equations, such as the summation of cubic polynomial melting point equations (Kleinhans et Mazur, 2007) or osmotic virial equations with cross coefficients (Prickett et al., 2011). These equations will not be developed in this paper.

These mathematical methods allowed the elaboration of ternary phase diagrams of saline solutions containing DMSO, glycerol, EG or propylene glycol (Kleinhans et Mazur, 2007; Benson et al., 2010; Prickett et al., 2011; Weng et al., 2011). The values obtained with these synthesised diagrams were in agreement with those reported in the literature and obtained with DSC (Figure 7).

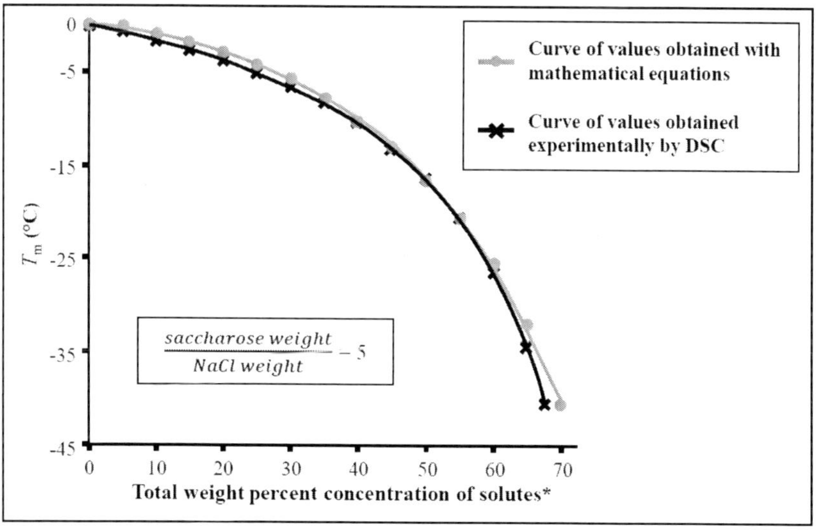

Figure 7. Partial ternary phase diagram for the system water, NaCl and saccharose, according to (Kleinhans et Mazur, 2007). *Mathematical methods have been developed to determine ternary or quaternary phase diagrams from binary phase diagrams, which are simpler to obtain. The synthesised diagrams obtained with these methods matched relatively well with those obtained experimentally with DSC. * solutes = saccharose + NaCl.*

The quaternary phase diagram of a system composed of water, NaCl, EG and saccharose has also been determined by comparing the two mathematical methods mentioned above (Han et al., 2010). Both methods were satisfactory for characterising this quaternary system. Nevertheless, the summation of cubic polynomial melting point equations seemed to be more accurate than the osmotic virial equations with cross coefficients to determine this quaternary phase diagram.

Characterisation of Intracellular Ice Formation

DSC can be used to obtain several types of crystallisation data. Consequently, this thermodynamic method has been used to better understand intracellular ice formation in living cells. Indeed, ice crystal formation is one of the most dangerous causes of cold damage (Vajta et Kuwayama, 2006). Particularly, intracellular ice, even in very small proportions, can induce significant cell damage and thus cellular death (Mullen et Critser, 2007). Consequently, a better understanding of intracellular ice formation can limit the cellular damages potentially induced by slow-freezing protocols. Three main recent studies have characterised intracellular ice formation with DSC.

The first study focused on intracellular ice formation in Xenopus oocytes (Kleinhans et al., 2006). A similar principle to the one developed for the optimisation of slow-freezing protocols was used (see part III of this chapter). In this study, cellular suspensions containing Xenopus oocytes were evaluated using five successive coolings and warmings applied in a differential scanning calorimeter:

1. A fast cooling was first applied until a temperature slightly higher than the extracellular crystallisation temperature.
2. A slow cooling was applied during extracellular ice formation. The extracellular crystallisation was induced by extracts of *Pseudomonas syringae*.
3. A slow warming was applied to reach a temperature just below the melting temperature in order to induce melting of the majority of the extracellular ice.
4. A slow cooling applied to several degrees allowed crystallisation of most of the aqueous solution.
5. A fast cooling (-10°C.min^{-1}) to -50°C allowed crystallisation of the intracellular ice.

The formations of extra- and intracellular ice crystals were partially separated in time. Consequently, intracellular crystallisation appeared as small peaks, which were superimposed on the extracellular crystallisation peak (main peak; Figure 8). The results obtained with cryomicroscopy confirmed the efficiency of this method for detecting intracellular crystallisation using DSC.

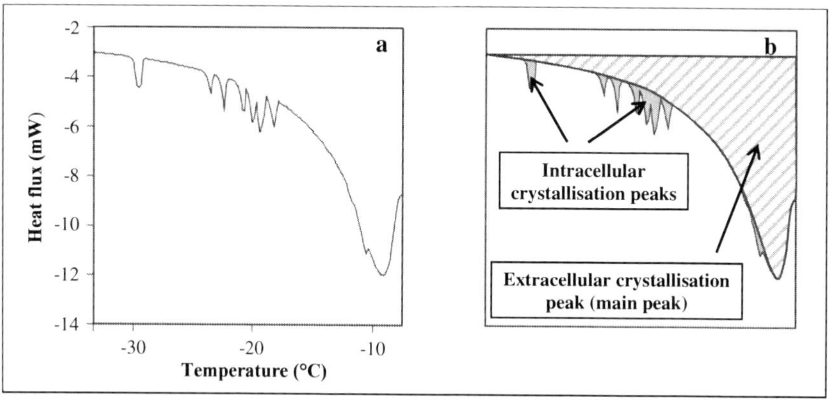

Figure 8. Thermogram (a) and explanatory scheme (b) allowing visualisation of the intracellular ice crystallisation in Xenopus oocytes, according to (Kleinhans et al., 2006). *Eight small peaks (in grey) are superimposed onto the main extracellular crystallisation peak (in strips). Each small peak corresponds to intracellular crystallisation occurring in one Xenopus oocyte.*

The intracellular crystallisation of late stage oocytes was unobservable or occurred at a high temperature. In regard to the unobservable crystallisation, the authors proposed that many late stage oocytes were partially damaged during extracellular crystallisation. Consequently, their intracellular freezing occurred during the extracellular freezing and was hidden. Otherwise, the presence of extracellular ice in close contact with cells is assumed to be an essential factor in the formation of intracellular ice (Mazur, 2004). Researchers also proposed that the large sizes of the late stage oocytes could increase the probability for ice-cell contact and, consequently, of intracellular crystallisation. This result could explain why the observed intracellular crystallisation occurred at a high temperature.

The intracellular crystallisation occurred at a low temperature in early stage oocytes. Consequently, the authors planned to use these early stage oocytes as models for understanding intracellular ice formation mechanisms. In particular, the introduction of aquaporins in their plasma membranes could allow verification of the inducing effect of extracellular ice on intracellular ice formation.

A second study focused on the impact of the cooling rate on the intracellular ice formation of *Saccharomyces cerevisiae* yeast cells (Seki et al., 2009). The influence of intracellular ice formation on cell survival was also evaluated. Intracellular ice formation was detected in several successive coolings and warmings at a constant rate.

Two types of solutions were evaluated: *(i)* distilled water and *(ii)* distilled water with yeast cells. In the two solutions, extracts of *Pseudomonas syringae* were used to minimise the supercooled state at the time of crystallisation. When yeast cells were added to distilled water, a "shoulder" was observed at the end of the crystallisation peak (Figure 9). This shoulder was not observed with distilled water alone and could be attributed to two possible phenomena:

- Heat released by extracellular crystallisation of water resulting from cell dehydration during the cooling.
- Heat released by intracellular crystallisation occurring in the yeast cells.

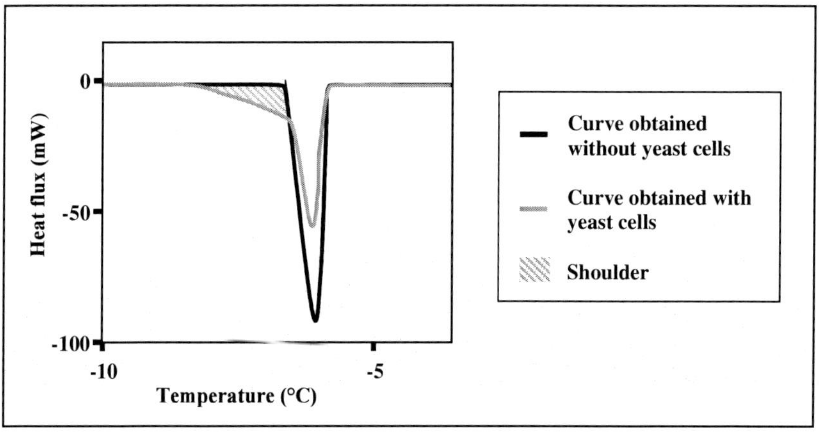

Figure 9. Thermogram allowing visualisation of the crystallisation occurring in distilled water with or without yeast cells, according to (Seki et al., 2009). *The shoulder (in stripes) does not appear during the characterisation of distilled water alone. Consequently, this result is due to the addition and influence of the cells. Two phenomena can explain the shoulder:*

- *Extracellular crystallisation following cellular dehydration.*
- *Intracellular crystallisation.*

The difference between the two phenomena cannot be realized on a single crystallisation peak. Nevertheless, there is extensive evidence that intracellular crystallisation causes membrane disruption. Consequently, after intracellular crystallisation, the intracellular media mixes with the extracellular media. Intracellular ice formation should thus cause a decrease of the shoulder area on successive cooling, and the amount of intracellular ice can be evaluated using this observed area decrease. When only extracellular ice forms, the cells remain osmotically active. As a consequence, the shoulder area should remain the same during the two successive coolings.

The authors observed that the shoulder decreased on successive coolings beyond a cooling rate of -20°C.min^{-1}, which allowed the identification of intracellular crystallisation. The amount of intracellular ice became more important as the cooling rate increased. As is usually observed with other cells, the survival rate of yeast cells depending on the cooling rate has the form of an inverted U. After comparison with other studies, the authors noticed that a cooling rate of -7°C.min^{-1} was correlated with a maximal survival rate and that cooling rates of -45°C.min^{-1} and -200°C.min^{-1} were associated with important decreases of the survival rates. This drop in viability occurs in the range of cooling rates in which intracellular crystallisation was identified by DSC. The authors concluded that intracellular ice formation was responsible for cellular death.

In a third study, the characterisation of the intracellular crystallisation was realised using the method developed for the optimisation of slow-freezing protocols (see part III of this chapter)(Mori et al., 2012). This study was conducted using cells suspensions of human dermal fibroblasts. When cells were cooled at -10°C.min^{-1}, -20°C.min^{-1} or -50°C.min^{-1}, a change in the gradient of the thermogram was observed during the cooling that allows the evaluation of Q_{initial}. This change was supposed to be due to intracellular crystallisation and was similar to that observed by Seki in the study described above (Figure 9). This hypothesis was confirmed by cryomicroscopy. As expected, no intracellular crystallisation was observed at a cooling rate of -0.5°C.min^{-1}, whereas cells cooled at -250°C.min^{-1} experienced mainly intracellular ice formation.

As explained before, a new thermal analysis method was applied and allowed the deconvolution of the curve of the heat released during the cooling three into three heat releases due to:*(i)* the extracellular ice formation, *(ii)* the water transport (during the cell dehydration) and *(iii)* the intracellular ice formation (Figure 6). Consequently, the intracellular crystallisation was evaluated. The onset temperatures of the intracellular crystallisation, as well as the fraction of cellular water converted to intracellular ice, were not always in agreement between the cryomicroscopy and the DSC measurements. These discrepancies could be due to the thermal lag of the DSC or the detection method of intracellular crystallisation with cryomicroscopy (darkening of the cell is not perfectly correlated with intracellular ice formation). However, the authors observed that the higher the cytocrit of the cell suspension was, the higher the onset temperature of the intracellular crystallisation was.

Finally, the authors evaluated the viability of cells after cooling them at -0.5°C.min^{-1}, -10°C.min^{-1}, -20°C.min^{-1} or -50°C.min^{-1}. As expected, they observed that the survival curve of the cells after freezing had the shape of an inverted U. The maximum viability was obtained with a cooling rate of -10°C.min^{-1}. Nevertheless, they noticed that a cooling rate of -10°C.min^{-1} is correlated with the crystallisation of 50% of the intracellular water. Despite the fact this phenomenon is surprising, 50% of intracellular crystallisation could consequently be optimal for this cell type.

Conclusion

DSC is a useful tool for studying and understanding slow-freezing protocols and crystallisation. Indeed, this thermodynamic method allows the characterisation of the behaviour of solutions and cell suspensions at negative temperatures. However, slow-freezing protocols are not the only protocols used for the cryopreservation of cells or tissues. In particular, another main cryopreservation method is vitrification. In vitrification protocols, the water does not crystallise, but is instead changed into a stable and solid amorphous state called glass. This vitrification phenomenon can be considered to be an extreme increase of the solution viscosity.

DSC can also be a useful tool to optimise the vitrification procedure. Indeed, DSC allows the evaluation (for a given solution) of the vitreous transition temperature and critical cooling and warming rates beyond which ice formation is avoided (Baudot, 1997). Consequently, DSC allows us to assess the tendency of a solution to form a glass as well as the stability of its amorphous state. DSC allowed the evaluation of the physical parameters of vitrification solutions for *in vitro*-produced porcine embryos (Men et al., 2011), murine oocytes (Seki et Mazur, 2009) and sheep ovaries (Baudot et al., 2007). In the last case, the critical cooling and warming rates were calculated for the vitrification solution alone and the vitrification solution impregnated into ovaries' fragments. The critical cooling rates were much higher when the solution was impregnated into ovary fragments. The authors deduced that the impregnation of the tissue with the solution was not sufficient. This study revealed the importance of evaluating the vitrification solutions in the presence of the cells or tissues.

From a more fundamental point of view, DSC has also already allowed the evaluation of aqueous solutions containing CPAs to evaluate their efficacies at forming glasses and selecting the most suitable CPA for vitrification. Several studies focused on the evaluation of solutions containing diethylformamide and dimethylformamide (Baudot et Boutron, 1998), propylene glycol and glycerol (Wowk et al., 1999), DMSO (Baudot et al., 2000), simple amino-alcohols (Baudot et al., 2002) and EG (Wowk et al., 1999; Baudot et Odagescu, 2004). Among the most used CPA, propylene glycol seems to have the best glass-forming tendency. In the same way, a study focused on the effect of structural modifications on the ability of a CPA to form a glass. Replacement of one hydroxyl group of polyols (EG, propylene glycol, etc.) by a methoxyl group decreased the critical cooling and warming rates and enhanced their abilities to form glass (Wowk et al., 1999). Consequently, methoxylated compounds could be used in vitrification solutions.

Acknowledgments

The authors would like to thank Prof. Peter Mazur and Prof. Ram V. Devireddy for allowing the use of their figures, with the agreement of Wiley and Elsevier publishers.

References

Alapati, R; Goff, K; Kubisch, HM; Devireddy, RV. Water transport in epididymal and ejaculated rhesus monkey (Macaca mulatta) sperm during freezing. *Cryobiology,* 2008 57, 182-185.

Alapati, R; Stout, M; Saenz, J; Gentry, GT, Jr.; Godke, RA, et al. Comparison of the permeability properties and post-thaw motility of ejaculated and epididymal bovine spermatozoa. *Cryobiology,* 2009 59, 164-170.

Baudot, A (1997)*Cryopreservation d'organes par vitrification Mesures calorimétriques et mesures diélectriques.* Grenoble: Institut National Polytechnique Grenoble. 240 p.

Baudot, A; Alger, L; Boutron, P. Glass-forming tendency in the system water-dimethyl sulfoxide. *Cryobiology,* 2000 40, 151-158.

Baudot, A; Boutron, P. Glass-forming tendency and stability of aqueous solutions of diethylformamide and dimethylformamide. *Cryobiology,* 1998 37, 187-199.

Baudot, A; Cacela, C; Duarte, ML; Fausto, R. Thermal study of simple amino-alcohol solutions. *Cryobiology,* 2002 44, 150-160.

Baudot, A; Courbiere, B; Odagescu, V; Salle, B; Mazoyer, C, et al. Towards whole sheep ovary cryopreservation. *Cryobiology,* 2007 55, 236-248.

Baudot, A; Odagescu, V. Thermal properties of ethylene glycol aqueous solutions. *Cryobiology,* 2004 48, 283-294.

Benson, JD; Bagchi, A; Han, X; Critser, JK; Woods, EJ. Melting point equations for the ternary system water/sodium chloride/ethylene glycol revisited. *Cryobiology,* 2010 61, 352-356.

Boutron, P. More accurate determination of the quantity of ice crystallized at low cooling rates in the glycerol and 1,2-propanediol aqueous solutions: comparison with equilibrium. *Cryobiology,* 1984 21, 183-191.

Bruyere, P; Baudot, A; Guyader-Joly, C; Guerin, P; Louis, G, et al. Improved cryopreservation of in vitro-produced bovine embryos using a chemically defined freezing medium. *Theriogenology,* 2012 78, 1294-1302.

Clas, SD; Dalton, CR; Hancock, BC. Differential scanning calorimetry: applications in drug development. *Pharm Sci Technol To,* 1999 2, 311-320.

Devireddy, RV; Campbell, WT; Buchanan, JT; Tiersch, TR. Freezing response of white bass (Morone chrysops) sperm cells. *Cryobiology,* 2006 52, 440-445.

Devireddy, RV; Coad, JE; Bischof, JC. Microscopic and calorimetric assessment of freezing processes in uterine fibroid tumor tissue. *Cryobiology,* 2001 42, 225-243.

Devireddy, RV; Fahrig, B; Godke, RA; Leibo, SP. Subzero water transport characteristics of boar spermatozoa confirm observed optimal cooling rates. *Mol Reprod Dev,* 2004 67, 446-457.

Devireddy, RV; Li, G; Leibo, SP. Suprazero cooling conditions significantly influence subzero permeability parameters of mammalian ovarian tissue. *Mol Reprod Dev,* 2006 73, 330-341.

Devireddy, RV; Raha, D; Bischof, JC. Measurement of water transport during freezing in cell suspensions using a differential scanning calorimeter. *Cryobiology,* 1998 36, 124-155.

Devireddy, RV; Swanlund, DJ; Olin, T; Vincente, W; Troedsson, MH, et al. Cryopreservation of equine sperm: optimal cooling rates in the presence and absence of cryoprotective agents determined using differential scanning calorimetry. *Biol Reprod,* 2002 66, 222-231.

Devireddy, RV; Swanlund, DJ; Roberts, KP; Bischof, JC. Subzero water permeability parameters of mouse spermatozoa in the presence of extracellular ice and cryoprotective agents. *Biol Reprod*, 1999 61, 764-775.

Devireddy, RV; Swanlund, DJ; Roberts, KP; Pryor, JL; Bischof, JC. The effect of extracellular ice and cryoprotective agents on the water permeability parameters of human sperm plasma membrane during freezing. *Hum Reprod,* 2000 15, 1125-1135.

Han, X; Liu, Y; Critser, JK. Determination of the quaternary phase diagram of the water-ethylene glycol-sucrose-NaCl system and a comparison between two theoretical methods for synthetic phase diagrams. *Cryobiology,* 2010 61, 52-57.

Jiang, Z; Imrie, CT; Hutchinson, JM. An introduction to temperature modulated differential scanning calorimetry (TMDSC): a relatively non-mathematical approach. *Thermochim Acta,* 2002 387, 75-93.

Karlsson, JO; Cravalho, EG; Toner, M. A model of diffusion-limited ice growth inside biological cells during freezing. *J Appl Phys,* 1994 75, 4442-4455.

Kleinhans, FW; Guenther, JF; Roberts, DM; Mazur, P. Analysis of intracellular ice nucleation in Xenopus oocytes by differential scanning calorimetry. *Cryobiology,* 2006 52, 128-138.

Kleinhans, FW; Mazur, P. Comparison of actual vs. synthesized ternary phase diagrams for solutes of cryobiological interest. *Cryobiology,* 2007 54, 212-222.

Li, G; Thirumala, S; Leibo, SP; Devireddy, RV. Subzero water transport characteristics and optimal rates of freezing rhesus monkey (*Macaca mulatta*) ovarian tissue. *Mol Reprod Dev,* 2006 73, 1600-1611.

Mazur, P. *Life in the frozen state: Principles of cryobiology.* Fuller BJ, lane N, Benson EE. Boca Raton: CRC Press; 2004, p.205-236.

Men, H; Zhao, C; Si, W; Murphy, CN; Spate, L, et al. Birth of piglets from *in vitro*-produced, zona-intact porcine embryos vitrified in a closed system. *Theriogenology,* 2011 76, 280-289.

Mori, S; Choi, J; Devireddy, RV; Bischof, JC. Calorimetric measurement of water transport and intracellular ice formation during freezing in cell suspensions. *Cryobiology,* 2012 65, 242-255.

Morris, GJ; Goodrich, M; Acton, E; Fonseca, F. The high viscosity encountered during freezing in glycerol solutions: effects on cryopreservation. *Cryobiology,* 2006 52, 323-334.

Mullen, SF; Critser, JK. *Oncofertility: Fertility Preservation for Cancer Survivors: The Science of Cryobiology.* Woodruff TK, Snyder KA. New York: Springer; 2007, p.83-103.

Pinisetty, D; Huang, C; Dong, Q; Tiersch, TR; Devireddy, RV. Subzero water permeability parameters and optimal freezing rates for sperm cells of the southern platyfish, Xiphophorus maculatus. *Cryobiology,* 2005 50, 250-263.

Prickett, RC; Elliott, JA; McGann, LE. Application of the multisolute osmotic virial equation to solutions containing electrolytes. *J Phys Chem B,* 2011 115, 14531-14543.

Salvetti, P; Joly, T; Baudot, A. Effect of antibiotics on thermodynamic properties of freezing media in rabbit species: A first calorimetric approach. *Cryobiology,* 2006 53, 268-275.

Seki, S; Kleinhans, FW; Mazur, P. Intracellular ice formation in yeast cells vs. cooling rate: predictions from modeling vs. experimental observations by differential scanning calorimetry. *Cryobiology,* 2009 58, 157-165.

Seki, S; Mazur, P. The dominance of warming rate over cooling rate in the survival of mouse oocytes subjected to a vitrification procedure. *Cryobiology,* 2009 59, 75-82.

Thirumala, S; Campbell, WT; Vicknair, MR; Tiersch, TR; Devireddy, RV. Freezing response and optimal cooling rates for cryopreserving sperm cells of striped bass, *Morone saxatilis. Theriogenology,* 2006 66, 964-973.

Thirumala, S; Ferrer, MS; Al-Jarrah, A; Eilts, BE; Paccamonti, DL, et al. Cryopreservation of canine spermatozoa: theoretical prediction of optimal cooling rates in the presence and absence of cryoprotective agents. *Cryobiology*, 2003 47, 109-124.

Thirumala, S; Gimble, JM; Devireddy, RV. Transport phenomena during freezing of adipose tissue derived adult stem cells. *Biotechnol Bioeng,* 2005 92, 372-383.

Thirumala, S; Huang, C; Dong, Q; Tiersch, TR; Devireddy, RV. A theoretically estimated optimal cooling rate for the cryopreservation of sperm cells from a live-bearing fish, the green swordtail Xiphophorus helleri. *Theriogenology,* 2005 63, 2395-2415.

Vajta, G; Kuwayama, M. Improving cryopreservation systems. *Theriogenology,* 2006 65, 236-244.

Weng, L; Li, W; Zuo, J. DSC determination of partial ternary phase diagrams of methanol/sodium chloride/water and propylene glycol/sodium chloride/water and their applications for synthesized diagrams. *Thermochim Acta,* 2011 512, 225-232.

Wowk, B; Darwin, M; Harris, SB; Russell, SR; Rasch, CM. Effects of solute methoxylation on glass-forming ability and stability of vitrification solutions. *Cryobiology,* 1999 39, 215-227.

In: Cryopreservation
Editors: A. Colvert and H. Coty
ISBN: 978-1-62618-474-9

Chapter 3

ALTERNATIVES TO CRYOPRESERVATION FOR THE SHORT AND LONG-TERM STORAGE OF MAMMALIAN CELLS

Karen Coopman*

Centre for Biological Engineering, Department of Chemical Engineering, Loughborough University, Loughborough, UK

Abstract

The ability to preserve mammalian cells has long been a critical part of cell-based research for several reasons. It allows for the transport of cells between laboratories or sites, ensures the availability of consistent starting material for the research through the establishment of cell banks and can uncouple cell-based assays from the culture process. Although cryopreservation, whether conventional slow freezing or vitrification, is widely used as a method for preserving cells long-term, it can result in low cell recovery post-thaw and the cryoprotective agents used in the freezing medium can be cytotoxic. With the emergence of a cell-based therapies industry, where clinical grade cells will need to be stored and transported, there is a growing need to develop scalable GMP-compatible preservation methods that retain not just high cell viability but also clinical efficacy. With these issues in mind, this review will explore two alternatives to cryopreservation: cell desiccation for long-term storage of cells and the short-term storage of cells under hypothermic (>0°C) conditions.

* E-mail address: k.coopman@lboro.ac.uk; Tel: +44 (0)1509 222513; Fax: +44 (0)1509 223923; Address: Centre for Biological Engineering, Department of Chemical Engineering, Loughborough University, Loughborough, LE11 3TU, UK.

Introduction

Cell-based therapies are part of a growing healthcare industry and are estimated to be worth $5 billion by 2014 [1].Such treatments have typically involved harvesting material such as bone marrow or islet cells from donors and transplantation or transfusion of freshly isolated or expanded cells. Cryopreservation, however, enables 'off the shelf' allogeneic therapies to be generated as the product, i.e. the cells, can be manufactured at a site physically separate from the clinic, stored and shipped as required. It also allows the generation of master and working banks of cells for allogeneic therapies and time for extensive QC and safety testing to be done before a product lot is released. Examples of cryopreserved cell therapies currently in clinical trials or use include CTX0E03 neural stem cells (ReNeuron Ltd, UK; [2]), Prochymal® (Osiris Therapeutics Inc, USA; [3]) and Dermagraft® (Advanced BioHealing, USA; [4]). Cryopreservation also underpins the growing business of cord blood banking. Despite its widespread use, there are several issues associated with the cryopreservation of cells and these will be briefly discussed before two alternatives to cryopreservation, namely cell desiccation and hypothermic storage, are explored.

Issues Associated with the Cryopreservation of Mammalian Cells

Although cryopreservation, whether conventional slow freezing or vitrification, is widely used as a method for preserving cells long-term, it can result in low cell recovery post-thaw. For instance, several studies reported poor survival and recovery rates of between 0 and 30% of human embryonic stem cell colonies or clumps when a slow freezing protocol was used [5-8]. Furthermore, although cell survival or viability is clearly an important measure of cryopreservation success, it is also important that cells retain their function long-term post-thaw, particularly when considering clinically relevant cells. For example, protocols have now been developed that enable cryopreservation of human hepatocytes from donor livers for cell transplantation. Although high numbers of viable cells can be recovered post-thaw, large proportions of cells lose their ability to attach to surfaces which is key during engraftment, and their metabolic functions are altered [9], indicating further improvements still need to be made.

Key to the success of any cryopreservation protocol is the use of cryoprotective agents (CPAs). One of the most commonly used CPAs is DMSO, despite the fact that it is reported to be toxic to cells [10] and has been linked to cell differentiation [11]. Cells therefore need to be processed rapidly and ideally at low temperatures. Although this may be practical on a small scale, this becomes more difficult when processing cells at an industrial scale, which could take upwards of an hour [12]. Furthermore, DMSO has also been reported to cause adverse reactions in patients [13] and so cells may need to be washed extensively prior to infusion or transplantation into patients. This extra processing step can results in significant cell loses [14], thus it is clear that protocols that avoid or minimize the use of DMSO must be considered, especially for preservation of cell-based therapies.

Given the large numbers of cells that may be required for allogeneic cell-based therapies [15] it is also important that cryopreservation protocols are developed that are scalable and GMP compliant. Although conventional slow freezing protocols are adaptable, vitrification is not a truly scalable process [10, 15, 16]. Vitrification also poses extra risks as cells in open cryocontainers are exposed directly to liquid nitrogen which can contain contaminants such as microorganisms [17]. With all frozen cells, whether conventionally frozen or vitrified, it is also important that they are stored appropriately, with no fluctuations in temperature [10, 18]. This means careful environmental control is required, especially during transport. Transport of vitrified cells can, for instance, be very expensive as dry shippers, which hold samples at -150°C without the risks associated with transporting liquid nitrogen, are required to prevent any partial thawing of the cells [10, 18].

Notably, a great deal of research is on-going to improve current cryopreservation protocols and overcome these issues. Some of these innovations have recently been reviewed elsewhere [10, 15, 19] but, for example, using BHK cells, the impact of prolonged exposure (up to 5 hours) to DMSO on cell recovery and therefore what the true processing window for banking large numbers of cells is, was determined [12]. The authors found that if using cryobags the banking should be done within 1.5-2 hours at room temperature. Furthermore, Hunt and colleagues found that $CD34^+$ cells could be exposed to 25% (w/w) DMSO for up to 1 hour at 2°C with no significant loss in the cells ability to generate colony forming units [20]. Although the authors only looked at exposure of freshly isolated cells to DMSO and not the impact that long-term exposure followed by freezing has on cell recovery, it is nonetheless encouraging to see that process conditions are starting to be defined beyond simply ‘work quickly’. The use of closed cryocontainers for vitrification to avoid direct exposure to liquid nitrogen is also being investigated [7, 21], as are 'bulk' vitrification methods [22], although even these are still relatively small scale [15]. Thus, although improvements are being made to the cryopreservation methods, it is worth exploring whether a preservation method, such as desiccation or hypothermic storage, which results in products that can be maintained at room temperature or in a refrigerated unit and could avoid the use of DMSO altogether, are viable alternatives.

Alternatives to Cryopreservation

Dry Preservation

One alternative to cryopreservation is to avoid low temperature storage and transport altogether by desiccating cells as dried formulations could be kept at room temperature. If feasible this would be both economically and practically advantageous, especially for long-term storage of cells. Although drying cells generally leads to membrane damage, protein denaturation and subsequent cell death, the inherent capacity of some organisms such as yeast cells and fungal spores to survive almost complete dehydration [23] suggests 'safe' dehydration and rehydration of mammalian cells should be possible. Here efforts to both freeze dry (lyophilise) and simply desiccate cells by, for instance, vacuum desiccation, will be reviewed.

Freeze drying, or lyophilisation, is typically used to preserve bacterial cells and fungi. Briefly, the process comprises three stages: freezing in order to partly crystallise the solvent so that it can be separated from the solutes, sublimation of the resultant ice (primary drying phase) and finally, a secondary drying phase where the majority of the remaining moisture is removed by desorption [23, 24]. Just as cryoprotectants are used in cryopreservation, cells can be pre-treated and indeed lyophilised in the presence of lyoprotectants, to minimize cell damage during the process. One compound that has been particularly researched is trehalose, a hydrophilic disaccharide which is known to accumulate in desiccation tolerant organisms [25].

The key challenge in using trehalose as a lyo- or indeed cryo-protectant is that it is membrane impermeable and thus achieving the intracellular concentrations of at least 100-200mM required for protection [26, 27] is non-trivial. Several different approaches have been reported including using fluid phase endocytosis [28], microinjection [29], mutant bacterial toxins as pore forming proteins [26, 27], genetically modifying cells to synthesise trehalose themselves [30] or express a trehalose transporter on their plasma membranes [31], osmotic shock [32, 33], electroporation [34] and a biopolymer (PP-50; [35]).How trehalose is currently understood to exert its protective effects has been thoroughly reviewed in Crowe et al. [25] and Ohtake and Wang [36] but includes its role as a water replacement inside cells and its effectiveness as a stable vitrification agent.

Efforts to lyophilise mammalian cells have largely focused on trying to generate freeze-dried blood, with some studies showing that erythrocytes and platelets can be successfully lyophilised, stored and rehydrated, although the processes are not yet optimal [28, 37-40].For instance, Goodrich et al. [37] reported that based on their patented method of erythrocyte lyophilisation, which involves lyophilising cells in the presence of monosaccharides such as glucose, cells maintained their metabolic function following rehydration although they note that it takes several days to reach the 1-2% residual moisture content in the samples that allows for storage at ambient temperatures. By chemically cross-linking cells in a reversible manner, Bakaltcheva and colleagues[39] were also able to successfully lyophilise erythrocytes, with less than 10% haemolysis reported in their optimal conditions. However, a final moisture content of 10-15% meant that cells had to be stored at 4°C, not ambient temperatures. At moisture contents of less than 10%, heamolysis upon rehydration increased significantly and the authors themselves raised concerns over the safety of transplanting these chemically treated cells into patients [39].Platelets have also been successfully lyophilised using trehalose as a lyoprotectant though it has been noted that the density at which cells are lyophilised and the method of how cells are rehydrated have a significant impact on the success rate [28]. Nonetheless, results are promising as a recent study using a model of deep arterial injury has even shown that lyophilised human platelets, when reconstituted into platelet free plasma, retained their haemostatic properties [40].

Although the genetic material of mouse embryonic stem cells, spermatozoa and sheep granulose cells has been preserved following freeze-drying [41-43], it was not until relatively recently that successful lyophilisation, storage for one week, and rehydration of nucleated cells was reported [44].The authors demonstrated viability of mononuclear cells isolated from human umbilical cord blood following rehydration, with similar levels of $CD34^+$haematopoietic stem cells present in the population before and after preservation, which were capable of differentiating into a number of cell types. In another study also focusing on haematopoietic progenitors, cells rehydrated after 4 week of storage at 25°C also retained their ability to generate colony forming units of different lineages. Rehydrated cells produced between 35-80% colony forming units relative to unprocessed controls, depending on the lineage studied [45].

For a number of therapies to be effective, cells may need to be encapsulated in order to, for instance, protect them from the patients' immune system [46].Therefore, the ability to freeze dry microencapsulated cells has also been investigated [47].The viability of human retinal pigment epithelial cells, either dispersed or microencapsulated, was determined before and after the drying step and, although it was clear that drying drastically reduced viability of cells compared to freezing alone, this proof of principal study indicates that encapsulated cells could be freeze-dried.

It is also possible to simply dry cells, under vacuum for instance, without the need for freezing. As with lyophilisation, the use of a protective agent such as trehalose appears to offer protection [32], although cell injury/death can still occur if the drying process is too slow or not uniform enough, thereby exposing cells to severe osmotic imbalances [48, 49].It has been shown that human mesenchymal stem cells (MSCs) exposed to trehalose and glycerol can be successfully dehydrated under vacuum and stored for 1 day [50].Upon rehydration they regained their normal morphology and adhesive capability, were >90% viable, proliferated and maintained expression of key surface markers. However, the authors did admit that there was a large amount of inconsistency from trial to trial and that longer storage times would need to be achieved. As they did not measure the levels of intracellular trehalose achieved it is possible that variation in this between runs or simply insufficient trehalose loading overall affected the success rate of desiccation. After expressing a trehalose transporter in CHO-cells, one group also reported that they could be successfully desiccated after loading with trehalose [31].Membrane integrity was retained in approximately 80% of cells when moisture content was reduced to as little as 2.6g water/g dry weight. However, upon rehydration these cells exhibited a vastly reduced proliferative capacity compared to control cells. Given that only ~23.5mM intracellular trehalose concentrations were achieved, well below the 100mM minimum mentioned earlier, it may be that with enhanced trehalose uptake the desiccation tolerance of CHO cells could also be improved.

Although these reports are promising, indicating that desiccation of cells for long-term storage may be possible in the future, there are several caveats. Clearly, as already mentioned, storage times must be increased. Secondly, if cells are to be used in cell therapy, it is important that cells are not simply viable immediate post-rehydration but retain normal function and clinical efficacy. For example, one study reported that human bone marrow-derived MSCs could be freeze-dried [51] but, whilst initial recovery following rehydration was high (~70%), cells were unable to adhere to tissue culture plastic as normal and died within 1 week. Similarly, Li and colleagues [52] successfully lyophilised Hep2G cells loaded with trehalose and modified to express two late embryogenesis abundant proteins which are associated with desiccation tolerance in anhydrobiotic animals and bacteria. A very high level of membrane integrity was retained up on rehydration (98%), but cell growth was decreased compared to non-dried controls. These studies highlight the need to look at the longer-term effects of these processes, not just immediate post-rehydration viability as the longer-term health of the cells may be compromised.

Based on their findings in human foreskin fibroblasts, Puhlev et al. [32] suggested that free-radical mediated damage may occur within desiccated cells. In order to achieve successful desiccation of nucleated cells we may therefore need to develop more complex protocols in order to overcome or prevent some of the stress that cells undergo during freezing and/or desiccation. For example, Chakraborty and colleagues have looked at ways to improve the drying process in order to reduce the osmotic stress that cells undergo. They developed a spin-drying technique [49] which enabled a much lower water content to be achieved (0.16g water/g dry weight) than in other studies, whilst retaining more than 95% membrane integrity of CHO cells. However, the authors do admit that this is preliminary data and the longer-term impact of such extreme desiccation on the CHO cells was not investigated. The impact of including additional lyoprotectants has also been explored. For instance, one study transfected HEK293 cells with a stress protein, p26 from *Artemia*, and found that it has a synergistic effect with trehalose in improving cell viability following air-drying and rehydration. An approximately 15% increase in membrane integrity was seen by the addition of the p26 as well as an increase in cell growth when cells had been dehydrated to 0.2 g water/g dry weight [53].

Although these studies are interesting and show that improvements to protocols can be made, when working with clinically relevant cell types it is critical to remember that the cell is the product and will end up in a patient. It will therefore be important to ensure that any additives used, or even the processes by which trehalose is loaded into cells, are accepted by the regulator. It is important to note that lyophilisation has been used to preserve bacterial cells [54] and its use in the pharmaceutical industry means that there is a precedent for lyophilisation at a large scale, making it an attractive option which should continue to be investigated.

Hypothermic Preservation

Another potential alternative to cryopreservation for mammalian cells is hypothermic preservation where cells are stored at low but not freezing temperatures, effectively pausing the cells by slowing down metabolism, but without the dangers of ice formation. At mild hypothermia (25-35°C) cells can still proliferate and grow and in fact, recombinant protein expression may be improved by culturing cells at reduced temperatures [55].More severe hypothermia and refrigerated storage (0-10°C) is needed to arrest growth and pause the cells but this would nonetheless be preferable to storing cells in liquid nitrogen and the use of dry shippers. This would be a short-term preservation method and indeed, it is already used in the pharmaceutical industry as it can uncouple cell culture from cell-based assays [56]. It allows researchers to store cells for several days if a piece of equipment is double-booked or breaks down, rather having to dispose of plates of cells, saving time and money. Just like cryopreservation, hypothermic preservation also uncouples production or harvest of cells for cell-based therapies from their clinical use. Although shelf life is more limited than with frozen products, this allows for an 'off the shelf' business model to exist, QC testing to be completed before cells are released and avoids the need for continuous culture.

Organs and blood fractions have long been stored at hypothermic temperatures prior to transplantation or transfusion [57, 58], however, cold injury can occur which can be amplified when cells are rewarmed. The changes that cells undergo during cold exposure include: decreased membrane fluidity, changes in pH, osmotic imbalances, activation of the unfolded protein response, oxidative stress [59, 59, 60] and often lead to the initiation of apoptosis. Therefore, several cold storage media have been designed to improve organ, tissue and cell preservation at temperatures in the range of 2-10°C.These include University of Wisconsin Solution (UWS, also known as Viaspan), EuroCollins, Celsior, Custodiol (also known as Bretschneider's HTK solution) and the HypoThermosol range [60-62]. Importantly, HypoThermosol-FRS (BioLife Solutions, USA) is a protein and serum free solution which is manufactured to cGMP standards so that it is suitable for use in cell therapy products. As the formation of reactive oxygen species is often implicated in cell injury, these solutions typically contain antioxidants such as glutathione [60, 63] as well as mimicking the ionic composition of the cells under cold temperatures, providing a pH buffering system and some nutrients. As an example the base formulation of HypoThermosol is given in Table 1 but it is notable that solutions containing iron chelators [60], apoptosis inhibitors [64] and even antifreeze proteins [65] have been tested.

Hypothermic preservation of a range of cell types has been investigated, including hepatocytes [60, 66], MSCs [67], red blood cells [58], neurons [68], renal cells [69] and endothelial cells [59, 62]. Despite the use of specially formulated preservation media, cell recovery from hypothermic storage can still be variable and is very dependent on the medium composition. For example, survival of human hepatocytes seeded in monolayer culture and stored at 4°C for up to 1 week in UWS was less than 20% [60].An optimized solution based on Custodiol and containing iron chelators, however, improved cell survival drastically such that cell death was less than 20% [60].Similarly, in a study using several cells types, Mathew and colleagues found that each was best preserved in a slightly different HypoThermosol variant [62].For instance, HypoThermosol supplemented with Trolox and RGD peptide best supported the survival of skeletal muscle cells stored in suspension at 4°C for 7 days whereas hepatocytes were best preserved in a formulation supplemented with a caspase inhibitor, FK041.

Hunt and colleagues [70] also demonstrated that it is possible to store cells in their normal culture medium. They reported that recombinant CHO and HEK293 cells (both adherent and suspension) could be paused for up to 3 weeks over a wide range of temperatures, from 4 to 24°C.Cells retained their capacity for growth and protein production once rewarmed to 37°C.Another study has shown that human embryonic stem cells in culture medium can also tolerate temperatures of 4 or 25°C for up to 48 hours with approximately 70% survival and retention of a normal karyotype and expression of pluripotency markers [71]. This does not seem to be universally applicable though as, in contrast, when human bone marrow derived MSCs were stored at 4°C in their growth medium they failed to survive, although they could be successfully stored for up to 4 days in HypoThermosol-FRS [67].Similar results were also obtained by Mathew et al. when they investigated the preservation of several different cell types [62].

It is important to note that, just as in the desiccation work, some studies have reported that although cell viability following hypothermic storage can be high, cell function can be negatively affected. For instance, storing isolated human hepatocytes in HypoThermosol-FRS at 4°C for up to 72 hours resulted in a cell viability of approximately 70% which was a drastic improvement over cells stored in UWS or the base HypoThermosolmedium. However, despite cell viability being high, cell metabolic function, was typically half that of freshly isolated cells [63].Another study which investigated whether bovine limbal epithelial cells could be stored at 4°C in calcium alginate hydrogels found that despite about 50% of cells extracted from the gels post-storage being viable, the colonies formed from these cells were noticeably smaller than those formed from cells that had not been stored [72]. These studies highlight that just as with other methods of preservation, it is important to not just focus on cell viability upon rewarming but also cell function and growth.

It is notable that many studies, including some of those referred to here, utilize adherent cells which have been seeded in plates, allowed to reach a certain degree of confluency and the growth medium is then replaced with hypothermic preservation solution before the plate is moved into cold storage. Although this is suitable for cells that will be used in plate-based cell assays such as those commonly used in pharmaceutical research, it is not suitable for storage of cell-based therapies. For example, a clinical dose of MSCs could exceed $2x10^8$ cells [73] and in our hands a confluent T75 contains approximately $3 \, x10^6$ MSCs [74] so a minimum of around 29 T175 flasks would be required for a single dose. Aside from the logistics of storing or transporting such large numbers of tissue culture flasks containing the liquid hypothermic preservation medium, it is also easier for the clinicians to receive vials or bags of cells that are ready for transplantation or transfusion, without needing to harvest cells from flasks. It is therefore important that anyone investigating the potential to store cells for cell-based therapies at 4°C does so under appropriate conditions.

Overall, hypothermic preservation does overcome some of the issues with cryopreservation. It avoids the use of DMSO, is amenable to scale up in that cells could be stored in vials or cryobags and GMP compliant preservation media is already available (Table 2).However, results so far have indicated that with some tweaking of the preservation solution, most cells are amenable to hypothermic storage. It will clearly always be a short-term solution compared to cryopreservation or indeed desiccation but, as is already apparent in organ transplantation and the blood banking industry, it has a role in getting cells from the site of harvest or production to the clinic. It remains to be seen how widely applicable hypothermic storage of cells is when the long-term effects on cell function and the impact of preservation in suspension are explored in more detail.

Conclusions and Future Perspectives

Although cryopreservation is by far the most common method of preserving mammalian cells it is not without its disadvantages. Here, two alternatives, summarized in Table 2, have been presented which could be used for either short or long-term preservation of cells. As hypothermic storage has been used in organ transplantation for so long, this route may be more acceptable to the cell biology community and regulators but desiccation should continue to be explored as an alternative to cryopreservation.

Table 1. Formulation of the base HypoThermosolmedium. It has an osmolality of 360 mOsm/kg. A vitamin E analogue, Trolox, is added to the base to generate HypoThermosol-FRS [62]

Component	Concentration
Sodium	100 mM
Potassium	42.5 mM
Magnesium	5 mM
Calcium	0.05 mM
Chlorine	17.1 mM
H_2PO4^-	10 mM
HCO_3^-	5 mM
HEPES	25 mM
Lactobionate	100 mM
Glucose	5 mM
Mannitol	20 mM
Sucrose	20 mM
Dextran-40	6%
Adenosine	2 mM
Glutathione	3 mM

Table 2. Summary of whether desiccation or hypothermic storage overcome any of the issues surrounding cryopreservation of mammalian cells.TBC; to be confirmed

	Hypothermic Storage	Desiccation
Maintains cell characteristics long-term	TBC	TBC
Widely applicable	Already used for organs and some cellular products	?
GMP compliant	amenable	amenable
Scalable	amenable	amenable
Avoids/limits use of DMSO	✓	✓
End-product is easily stored and/or transported	✓	✓

As these methods evolve it will also be interesting to see whether, in order to extend the shelf life of products, these preservation methods are combined in the future. For instance, one can envisage a scenario where cells are frozen in the first instance for long-term storage, then thawed, washed and resuspended in hypothermic storage medium for transport to the clinic as and when needed. How cells will respond to these multiple stresses remains to be seen.

References

[1] Mason, C.; Brindley, D.A.; Culme-Seymour, E.J.; Davie, N.L. Cell therapy industry: billion dollar global business with unlimited potential. *Regen. Med.* 2011, 6, 265-272.

[2] Stroemer, P.; Patel, S.; Hope, A.; Oliveira, C.; Pollock, K.; Sinden, J. The neural stem cell line CTX0E03 promotes behavioral recovery and endogenous neurogenesis after experimental stroke in a dose-dependent fashion. Neurorehabil. *Neural Repair* 2009, 23, 895-909.

[3] Kebriaei, P.; Isola, L.; Bahceci, E.; Holland, K.; Rowley, S.; McGuirk, J.; Devetten, M.; Jansen, J.; Herzig, R.; Schuster, M.; Monroy, R.; Uberti, J. Adult human mesenchymal stem cells added to corticosteroid therapy for the treatment of acute graft-versus-host disease. *Biol. Blood Marrow Transplant.* 2009, 15, 804-811.

[4] Mason, C. and Manzotti, E. Regenerative medicine cell therapies: numbers of units manufactured and patients treated between 1988 and 2010. *Regen. Med.* 2010, 5, 307-313.

[5] Hunt, C.J. and Timmons, P.M. Cryopreservation of human embryonic stem cell lines. *Methods Mol. Biol.* 2007, 368, 261-270.

[6] Li, Y.; Tan, J.C.; Li, L.S. Comparison of three methods for cryopreservation of human embryonic stem cells. *Fertil. Steril.* 2010, 93, 999-1005.

[7] Richards, M.; Fong, C.Y.; Tan, S.; Chan, W.K.; Bongso, A. An efficient and safe xeno-free cryopreservation method for the storage of human embryonic stem cells. *Stem Cells* 2004, 22, 779-789.

[8] Xu, X.; Cowley, S.; Flaim, C.J.; James, W.; Seymour, L.; Cui, Z. The roles of apoptotic pathways in the low recovery rate after cryopreservation of dissociated human embryonic stem cells. *Biotechnol. Prog.* 2010,

[9] Terry, C.; Dhawan, A.; Mitry, R.R.; Hughes, R.D. Cryopreservation of isolated human hepatocytes for transplantation: State of the art. *Cryobiology* 2006, 53, 149-159.

[10] Hunt, C.J. Cryopreservation of Human Stem Cells for Clinical Application: A Review. *Transfus. Med. Hemother* 2011, 38, 107-123.

[11] Santos, N.C.; Figueira-Coelho, J.; Martins-Silva, J.; Saldanha, C. Multidisciplinary utilization of dimethyl sulfoxide: pharmacological, cellular, and molecular aspects. *Biochem. Pharmacol.* 2003, 65, 1035-1041.

[12] Heidemann, R.; Lunse, S.; Tran, D.; Zhang, C. Characterization of cell-banking parameters for the cryopreservation of mammalian cell lines in 100-mL cryobags. *Biotechnol. Prog.* 2010, 26, 1154-1163.

[13] Cox, M.A.; Kastrup, J.; Hrubisko, M. Historical perspectives and the future of adverse reactions associated with haemopoietic stem cells cryopreserved with dimethyl sulfoxide. *Cell.Tissue Bank.* 2012, 13, 203-215.

[14] Stolzing, A.; Naaldijk, Y.; Fedorova, V.; Sethe, S. Hydroxyethylstarch in cryopreservation - mechanisms, benefits and problems. *Transfus. Apher. Sci.* 2012, 46, 137-147.

[15] Coopman, K. Large-scale compatible methods for the preservation of human embryonic stem cells: current perspectives. *Biotechnol. Prog.* 2011, 27, 1511-1521.

[16] Unger, C.; Skottman, H.; Blomberg, P.; Dilber, M.S.; Hovatta, O. Good manufacturing practice and clinical-grade human embryonic stem cell lines. *Hum. Mol. Genet.* 2008, 17, R48-53.

[17] Mirabet, V.; Alvarez, M.; Solves, P.; Ocete, D.; Gimeno, C. Use of liquid nitrogen during storage in a cell and tissue bank: contamination risk and effect on the detectability of potential viral contaminants. *Cryobiology* 2012, 64, 121-123.

[18] Thirumala, S.; Goebel, W.S.; Woods, E.J. Clinical grade adult stem cell banking. *Organogenesis* 2009, 5, 143-154.

[19] Li, Y. and Ma, T. Bioprocessing of Cryopreservation for Large-Scale Banking of Human Pluripotent Stem Cells. *BioResearch Open Access* 2012, 1, 205-214.

[20] Hunt, C.J.; Armitage, S.E.; Pegg, D.E. Cryopreservation of umbilical cord blood: 2. Tolerance of CD34(+) cells to multimolar dimethyl sulphoxide and the effect of cooling rate on recovery after freezing and thawing. *Cryobiology* 2003, 46, 76-87.

[21] Kuleshova, L.L.; Tan, F.C.; Magalhaes, R.; Gouk, S.S.; Lee, K.H.; Dawe, G.S. Effective cryopreservation of neural stem or progenitor cells without serum or proteins by vitrification. *Cell Transplant.* 2009, 18, 135-144.

[22] Li, T.; Mai, Q.; Gao, J.; Zhou, C. Cryopreservation of human embryonic stem cells with a new bulk vitrification method. *Biol. Reprod.* 2010, 82, 848-853.

[23] Wolkers, W.F.; Tablin, F.; Crowe, J.H. From anhydrobiosis to freeze-drying of eukaryotic cells. *Comp. Biochem. Physiol. A. Mol. Integr. Physiol.* 2002, 131, 535-543.

[24] Sundaramurthi, P. and Suryanarayanan, R. Calorimetry and complementary techniques to characterize frozen and freeze-dried systems. *Adv. Drug Deliv. Rev.* 2012, 64, 384-395.

[25] Crowe, J.H.; Crowe, L.M.; Wolkers, W.F.; Oliver, A.E.; Ma, X.; Auh, J.H.; Tang, M.; Zhu, S.; Norris, J.; Tablin, F. Stabilization of dry Mammalian cells: lessons from nature. *Integr. Comp. Biol.* 2005, 45, 810-820.

[26] Eroglu, A.; Russo, M.J.; Bieganski, R.; Fowler, A.; Cheley, S.; Bayley, H.; Toner, M. Intracellular trehalose improves the survival of cryopreserved mammalian cells. *Nat. Biotechnol.* 2000, 18, 163-167.

[27] Chen, T.; Acker, J.P.; Eroglu, A.; Cheley, S.; Bayley, H.; Fowler, A.; Toner, M. Beneficial effect of intracellular trehalose on the membrane integrity of dried mammalian cells. *Cryobiology* 2001, 43, 168-181.

[28] Wolkers, W.F.; Walker, N.J.; Tablin, F.; Crowe, J.H. Human platelets loaded with trehalose survive freeze-drying. *Cryobiology* 2001, 42, 79-87.

[29] Eroglu, A.; Toner, M.; Toth, T.L. Beneficial effect of microinjected trehalose on the cryosurvival of human oocytes. *Fertil. Steril.* 2002, 77, 152-158.

[30] Guo, N.; Puhlev, I.; Brown, D.R.; Mansbridge, J.; Levine, F. Trehalose expression confers desiccation tolerance on human cells. *Nat. Biotechnol.* 2000, 18, 168-171.

[31] Chakraborty, N.; Menze, M.A.; Elmoazzen, H.; Vu, H.; Yarmush, M.L.; Hand, S.C.; Toner, M. Trehalose transporter from African chironomid larvae improves desiccation tolerance of Chinese hamster ovary cells. *Cryobiology* 2012, 64, 91-96.

[32] Puhlev, I.; Guo, N.; Brown, D.R.; Levine, F. Desiccation tolerance in human cells. *Cryobiology* 2001, 42, 207-217.

[33] Satpathy, G.R.; Torok, Z.; Bali, R.; Dwyre, D.M.; Little, E.; Walker, N.J.; Tablin, F.; Crowe, J.H.; Tsvetkova, N.M. Loading red blood cells with trehalose: a step towards biostabilization. *Cryobiology* 2004, 49, 123-136.

[34] Zhou, X.; Yuan, J.; Liu, J.; Liu, B. Loading trehalose into red blood cells by electroporation and its application in freeze-drying. *Cryo Letters* 2010, 31, 147-156.

[35] Lynch, A.L.; Chen, R.; Dominowski, P.J.; Shalaev, E.Y.; Yancey, R.J., Jr; Slater, N.K. Biopolymer mediated trehalose uptake for enhanced erythrocyte cryosurvival. *Biomaterials* 2010, 31, 6096-6103.

[36] Ohtake, S. and Wang, Y.J. Trehalose: current use and future applications. *J. Pharm. Sci.* 2011, 100, 2020-2053.

[37] Goodrich, R.P.; Sowemimo-Coker, S.O.; Zerez, C.R.; Tanaka, K.R. Preservation of metabolic activity in lyophilized human erythrocytes. *Proc. Natl. Acad. Sci. U. S. A.* 1992, 89, 967-971.

[38] Han, Y.; Quan, G.B.; Liu, X.Z.; Ma, E.P.; Liu, A.; Jin, P.; Cao, W. Improved preservation of human red blood cells by lyophilization. *Cryobiology* 2005, 51, 152-164.

[39] Bakaltcheva, I.; Leslie, S.; MacDonald, V.; Spargo, B.; Rudolph, A. Reversible cross-linking and CO treatment as an approach in red cell stabilization. *Cryobiology* 2000, 40, 343-359.

[40] Joshi, N.V.; Raftis, J.B.; Lucking, A.J.; Hunter, A.H.; Millar, M.; Fitzpatrick, M.; Feuerstein, G.Z.; Newby, D.E. Lyophilised reconstituted human platelets increase thrombus formation in a clinical ex vivo model of deep arterial injury. *Thromb. Haemost.* 2012, 108, 176-182.

[41] Ono, T.; Mizutani, E.; Li, C.; Wakayama, T. Nuclear transfer preserves the nuclear genome of freeze-dried mouse cells. *J. Reprod. Dev.* 2008, 54, 486-491.

[42] Loi, P.; Matzukawa, K.; Ptak, G.; Natan, Y.; Fulka, J.,Jr; Arav, A. Nuclear transfer of freeze-dried somatic cells into enucleated sheep oocytes. *Reprod. Domest. Anim.* 2008, 43 Suppl 2, 417-422.

[43] Wakayama, T. and Yanagimachi, R. Development of normal mice from oocytes injected with freeze-dried spermatozoa. *Nat. Biotechnol.* 1998, 16, 639-641.

[44] Natan, D.; Nagler, A.; Arav, A. Freeze-drying of mononuclear cells derived from umbilical cord blood followed by colony formation. *PLoS One* 2009, 4, e5240.

[45] Buchanan, S.S.; Pyatt, D.W.; Carpenter, J.F. Preservation of differentiation and clonogenic potential of human hematopoietic stem and progenitor cells during lyophilization and ambient storage. *PLoS One* 2010, 5, 10.1371/journal.pone.0012518.

[46] Krishnamurthy, N.V. and Gimi, B. Encapsulated cell grafts to treat cellular deficiencies and dysfunction. *Crit. Rev. Biomed. Eng.* 2011, 39, 473-491.

[47] Wikstrom, J.; Elomaa, M.; Nevala, L.; Raikkonen, J.; Heljo, P.; Urtti, A.; Yliperttula, M. Viability of freeze dried microencapsulated human retinal pigment epithelial cells. *Eur. J. Pharm. Sci.* 2012, 47, 520-526.

[48] Chakraborty, N.; Biswas, D.; Parker, W.; Moyer, P.; Elliott, G.D. A role for microwave processing in the dry preservation of mammalian cells. *Biotechnol. Bioeng.* 2008, 100, 782-796.

[49] Chakraborty, N.; Chang, A.; Elmoazzen, H.; Menze, M.A.; Hand, S.C.; Toner, M. A spin-drying technique for lyopreservation of mammalian cells. *Ann. Biomed. Eng.* 2011, 39, 1582-1591.

[50] Gordon, S.L.; Oppenheimer, S.R.; Mackay, A.M.; Brunnabend, J.; Puhlev, I.; Levine, F. Recovery of human mesenchymal stem cells following dehydration and rehydration. *Cryobiology* 2001, 43, 182-187.

[51] Zhang, S.; Qian, H.; Wang, Z.; Fan, J.; Zhou, Q.; Chen, G.; Rui, L.; Fu, S.; Sun, J. Preliminary study on the freeze-drying of human bone marrow-derived mesenchymal stem cells. *Journal of Zhejiang University-Science* B 2010, 11, 889-894.

[52] Li, S.; Chakraborty, N.; Borcar, A.; Menze, M.A.; Toner, M.; Hand, S.C. Late embryogenesis abundant proteins protect human hepatoma cells during acute desiccation. *Proc. Natl. Acad.* Sci. U. S. A. 2012, 109, 20859-20864.

[53] Ma, X.; Jamil, K.; Macrae, T.H.; Clegg, J.S.; Russell, J.M.; Villeneuve, T.S.; Euloth, M.; Sun, Y.; Crowe, J.H.; Tablin, F.; Oliver, A.E. A small stress protein acts synergistically with trehalose to confer desiccation tolerance on mammalian cells. *Cryobiology* 2005, 51, 15-28.

[54] Day, J.G. and Stacey, G.N. Biobanking. *Mol. Biotechnol.* 2008, 40, 202-213.

[55] Roobol, A.; Carden, M.J.; Newsam, R.J.; Smales, C.M. Biochemical insights into the mechanisms central to the response of mammalian cells to cold stress and subsequent rewarming. *FEBS J.* 2009, 276, 286-302.

[56] Wise, H.; Abel, P.W.; Cawkill, D. Use of reduced temperature cell pausing to enhance flexibility of cell-based assays. *J. Biomol. Screen.* 2009, 14, 716-722.

[57] Fowler, A. and Toner, M. Cryo-injury and biopreservation. *Ann. N. Y. Acad. Sci.* 2005, 1066, 119-135.

[58] Holovati, J.L. and Acker, J.P. Emerging Role for Use of Liposomes in the Biopreservation of Red Blood Cells. *Transfus. Med. Hemother* 2011, 38, 99-106.

[59] Corwin, W.L.; Baust, J.M.; Baust, J.G.; Van Buskirk, R.G. The unfolded protein response in human corneal endothelial cells following hypothermic storage: implications of a novel stress pathway. *Cryobiology* 2011, 63, 46-55.

[60] Pless, G.; Sauer, I.M.; Rauen, U. Improvement of the cold storage of isolated human hepatocytes. *Cell Transplant.* 2012, 21, 23-37.

[61] Van Buskirk, R.G.; Baust, J.M.; Snyder, K.K.; Mathew, A.J.; Baust, J.G. Hypothermic Storage and Cryopreservation: Successful Short- and Long-

Term Preservation of Cells and Tissues. *BioProcess International* 2004, 2, 42-49.

[62] Mathew, A.J.; Baust, J.M.; Van Buskirk, R.G.; Baust, J.G. Cell preservation in reparative and regenerative medicine: evolution of individualized solution composition. *Tissue Eng.* 2004, 10, 1662-1671.

[63] Ostrowska, A.; Gu, K.; Bode, D.C.; Van Buskirk, R.G. Hypothermic storage of isolated human hepatocytes: a comparison between University of Wisconsin solution and a hypothermosol platform. *Arch. Toxicol.* 2009, 83, 493-502.

[64] Baust, J.M. Advances in Media for Cryopreservation and Hypothermic Storage. *BioProcess International* 2005, 3, 46-56.

[65] Hirano, Y.; Nishimiya, Y.; Matsumoto, S.; Matsushita, M.; Todo, S.; Miura, A.; Komatsu, Y.; Tsuda, S. Hypothermic preservation effect on mammalian cells of type III antifreeze proteins from notched-fin eelpout. *Cryobiology* 2008, 57, 46-51.

[66] Meng, Q. Hypothermic preservation of hepatocytes. *Biotechnol. Prog.* 2003, 19, 1118-1127.

[67] Ginis, I.; Grinblat, B.; Shirvan, M.H. Evaluation of bone marrow-derived mesenchymal stem cells after cryopreservation and hypothermic storage in clinically safe medium. *Tissue Eng. Part C. Methods* 2012, 18, 453-463.

[68] Nikkhah, G.; Eberhard, J.; Olsson, M.; Bjorklund, A. Preservation of fetal ventral mesencephalic cells by cool storage: in-vitro viability and TH-positive neuron survival after microtransplantation to the striatum. *Brain Res.* 1995, 687, 22-34.

[69] Mathew, A.J.; Van Buskirk, R.G.; Baust, J.G. Improved Hypothermic Preservation of Human Renal Cells Through Suppression of Both Apoptosis and Necrosis. *Cell Preservation Technology* 2004, 1, 239-253.

[70] Hunt, L.; Hacker, D.L.; Grosjean, F.; De Jesus, M.; Uebersax, L.; Jordan, M.; Wurm, F.M. Low-temperature pausing of cultivated mammalian cells. *Biotechnol. Bioeng.* 2005, 89, 157-163.

[71] Heng, B.C.; Vinoth, K.J.; Liu, H.; Hande, M.P.; Cao, T. Low temperature tolerance of human embryonic stem cells. *Int. J. Med. Sci.* 2006, 3, 124-129.

[72] Wright, B.; Cave, R.A.; Cook, J.P.; Khutoryanskiy, V.V.; Mi, S.; Chen, B.; Leyland, M.; Connon, C.J. Enhanced viability of corneal epithelial cells for efficient transport/storage using a structurally modified calcium alginate hydrogel. *Regen. Med.* 2012, 7, 295-307.

[73] Jung, S.; Panchalingam, K.M.; Wuerth, R.D.; Rosenberg, L.; Behie, L.A. Large-scale production of human mesenchymal stem cells for clinical applications. *Biotechnology and Applied Biochemistry* 2012, 59, 106-120.

[74] Rafiq, QA.; Coopman, K.; Nienow, AW.; Hewitt, CJ. A quantitative approach for understanding small-scale human mesenchymal stem cell culture – implications for large-scale bioprocess development. *Biotechnol. J.* 2013 DOI 10.1002/biot.201200197.

In: Cryopreservation
Editors: A. Colvert and H. Coty

ISBN: 978-1-62618-474-9

Chapter 4

Mathematical Modeling and Simulations in Cryopreservation of Living Cells

N. D. Botkin*, K.- H. Hoffmann and V. L. Turova
Technische Universität München, Mathematics Centre,
Chair of Mathematical Modelling, Garching b. Munich, Germany

Abstract

Cryopreservation of living cells is a necessary part of many medical procedures such as organ and tissue transplantation, conservation of reproductive and stem cells, etc. However, cells and tissues can be damaged during their freezing and subsequent thawing. Because of complexity of physical and chemical processes, mathematical models are necessary for finding optimal control regimes which increase the survival rate of cells. In this chapter, a phenomenological mathematical model based on mean values of physical quantities and a distributed phase field model of ice formation are described, and new methods of their investigation are developed. The results are applied to the design and implementation of optimized cooling protocols in a real freezing plant. The objective of modeling is to reduce cell damage caused, e.g. by sudden temperature drop due to supercooling, by dehydration and shrinkage due to the osmotic flow through cell membranes occurring at slow cooling rates, or by the growth of dendrite seeds that can cause formation of large ice crystals

*E-mail address: botkin@ma.tum.de

during subsequent thawing. A dynamic programming approach is used as a tool for computing optimized cooling regimes. These techniques are based on effective numerical methods for solving related Hamilton-Jacobi equations. Numerous numerical simulations aimed to the optimization of cooling and thawing protocols accounting for supercooling effects, dendritic growth of ice crystals, and cell shape changes due to osmotic effects are presented.

PACS: 87.85.Tu, 02.30Yy, 02.70.Bf

Keywords: freezing and thawing of biological cells, cooling and warming protocols, damaging factors, mathematical model, optimal control problem, value function, finite-difference scheme

AMS Subject Classification: 49N90, 49L20, 49L25, 34K28

1. Introduction

1.1. Damaging Factors of Cryopreservation

Cryopreservation of living cells and tissues has two related aspects. First, cells should be frozen so that the chemical reactions of life are suspended for a long duration. Second, frozen cells have to be warmed and thawed before they can be used. It is stated in many experiments that beneficial thawing conditions depend on prior cooling rates and cryoprotective agents used during freezing. There are many damaging effects and relations between them. Some of those are outlined below.

A sudden temperature drop or rise can cause thermal shock damage that begins at the plasma membrane as a result of differential shrinkage of various membrane components, mechanical shearing, and changes in membrane topology.

Another damaging factor is an increase in the concentration of salt in the extracellular solution, which leads to cellular dehydration and shrinkage due to the osmotic flow through the cell membrane. For every cell type, there exists a dehydration threshold beyond which cellular structures suffer irreversible damage. Such an effect occurs, if cooling is too slow. On the other hand, slow warming causes strong dilution of the extracellular solute, which leads to rehydration and swelling of cells.

The main factor which causes loss of viability is the formation of intracellular ice crystals. If cooling is rapid, water inside cells forms very small ice nuclei, whose aggregation in large irregular crystals (dendrites) during subsequent thawing can damage cells. It should be noted that the growth of dendrites is very unstable so that any change of physicochemical conditions may affect this process very dramatically. Mathematical description of such phenomena is a challenging task, both conceptually and computationally.

The highest survival rates are obtained at a range of cooling rates referred to as the transition zone in which the combined effect of the above mentioned mechanisms is at a minimum. Notice that the rate of thawing is closely associated with the rate of freezing. Generally, rapid thawing is preferable to slow one. Although the rate of thawing does not have a great effect on cells which were frozen slowly, it strongly effects on cells that were frozen more rapidly. These observations show that intracellular damages caused by ice occur during thawing but not when the crystals are initially formed at freezing stage. There are arguments that better results can be obtained, if the temperature falls not monotonically in time, especially in the range were the latent heat is released. Thus, one can speak about optimal cooling protocols.

Another control tool is cryoprotective agents which vary eutectic properties of solutions. Cryoprotective agents work as follows. By lowering the freezing point of the extracellular fluid, they reduce the efflux of intracellular water and therefore prevent cells from shrinking to its critical volume. Thus, cryoprotectors shift the entire transition zone in the direction of lower rates.

1.2. Mathematical Models in Cryobiology

Mathematical modeling in cryobiology is basically developed for freezing of living cells. First of all, the questions of intracellular ice formation and dehydration of cells have been studied. Exemplary, the following methods and models in this topic can be mentioned.

A physicochemical model based on the modified classical heterogeneous nucleation theory was proposed in [1] to analyze ice formation inside biological cells during freezing. According to this model, intracellular ice formation may be catalyzed either by the plasma cell membrane subjected to the influence of the external ice or by intracellular particles depending on freezing conditions. The model for intracellular ice formation is coupled with a model describing the kinetics of water transport to predict the thermodynamic state of the cytoplasm.

A method to estimate the nucleation frequency was proposed. It was asserted that the uncertainties in the experimental conditions can be minimized, if cells are cooled at rapid rates to prevent water transport during freezing. The model was applied to the cryopreservation of mouse oocytes.

A three-part coupled model of cell dehydration, nucleation, and crystal growth is used in [2] to study intracellular ice formation (IIF) in cultured hepatocytes frozen in the presence of dimethyl sulfoxide (DMSO). The heterogeneous nucleation temperature predicted as a function of DMSO concentration is in a good agreement with experimental data. Simulations predict and explain experimentally observed effects of cooling rate, warming rate, and storage temperature on hepatocytes. It was observed that no IIF occurs for cooling rates less than 10 degrees C/min. The predicted volume of intracellular ice increases with increasing cooling rate. Cell damage correlates positively with predicted intracellular ice volume, and an upper limit for the critical ice content is estimated to be 3.7% of the isotonic water content.

Paper [3] studies the effect of cryoprotectors (DMSO and glycerol) on intracellular ice formation using small- and wide-angle X-ray scattering. This effect decreases as the cryoprotector concentration increases. The formation of ice does not influence the membrane structure at all, if cryoprotector mole fractions is higher than a certain threshold.

The work [4] and the related work [5] study a phase-field model that describes effects associated with supercooling. The model contains a phase-field parameter that varies from 0 (solid) to 1 (liquid) and defines the freezing front. Parameters such as surface tension and interface kinetic coefficients are included in the model in order to capture the physics of the problem. In addition, conditions for nucleation are considered and modeled. This study provides a better understanding of how supercooling impacts biological cells and tissues and how to prevent damage of them.

The works [6] and [7] are comprehensive contributions to a very interesting book "Life in the Frozen State". This book describes the current state of knowledge of how living cells and complex organisms survive freezing to very low temperatures. These contributions discuss both biological and physicochemical principles of cryopreservation in all their complexity. The most injuring effects of freezing and thawing are discussed from the viewpoint of modeling and measuring. The contributions are closely related to the works [1] and [2].

Paper [8] studies intracellular ice formation (IIF) for cells cultured in micropatterned linear constructs. By fitting a Markov chain model to measured

IIF data for micropatterned cell pairs, the kinetics of IIF in linear four-cell constructs was predicted. Moreover, the model can be extended to predict IIF in larger tissues using Monte Carlo techniques.

Dissertation [9] contains numerous experimental data related to freezing and thawing of tissues. This includes data as well taken from literature as those measured by the author. For example, cell membrane permeabilities for murine insulinomas for water and different cryopreservation agents (CPA) are determined, and CPA transport through diverse extracellular matrices is measured. Methods of microscopy control of the viability, the fresh level, and the level of pluripotency of frozen-thawed cells are developed. Simple mathematical balance models operating with scalar values and ignoring phase transitions are used to quantify effects of cooling and thawing.

2. Freezing of Small Tissue Samples

Usually, tissue samples are frozen using special plants e.g. of the IceCube family developed by SY-LAB, Geräte GmbH (Austria). The main part of such plants is a freezing chamber equipped with a cooling system. Plastic ampoules with tissue samples surrounded by liquid (see Figure 1) are placed into the chamber. The plant is controlled by a computer that allows the user to prescribe cooling protocols to be tracked.

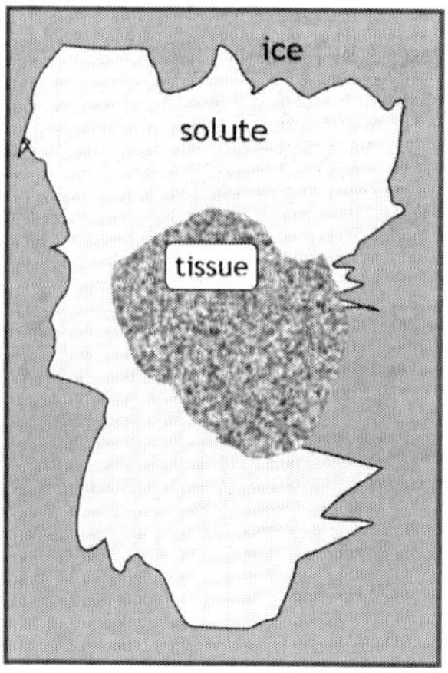

Figure 1. Ampoule with a tissue sample and a surrounding solute.

This section is devoted to the statement of mathematical models describing the process of freezing and to the application of optimal control theory to the design of optimal cooling protocols that minimize damaging effects caused by the release of the latent heat and by delayed freezing of intracellular water.

The first model utilizes mean values of thermodynamical parameters to describe the mean temperature of the ampoule. The control here is the temperature regime in the freezing chamber. A version of such a model along with the design of improved cooling protocols is implemented and tested in experiments on freezing of dental tissue samples.

The second model of freezing process deals with spatially distributed parameters and describes ice formation in the liquid surrounding the tissue. This approach is based on the so-called phase-field models governed by partial differential equations. They were introduced by Caginalp (see [10]) and studied by many scientists. We base the design of optimized cooling protocols on results of [11] where an optimal control problem for a phase-field model is stated and investigated from the mathematical and algorithmic points of view. Optimal controls are computed using the gradient descent method based on adjoint equations.

The third model should describe ice formation inside the extracellular matrix of the tissue. This includes phase changes in the extracellular liquid confined inside of small pores and mechanical stresses exerted on cell boundaries due to delayed freezing of the intracellular water.

2.1. Mean Value Temperature Response Model

Freezer IceCube 15M (SY-LAB, Geräte GmbH, Austria) is developed for controlled freezing tissue samples contained in plastic ampoules (see Figure 2 to the right). The main part of the plant is a freezing chamber equipped with a cooling system based on gas nitrogen, a rack for placing ampoules, and two temperature sensors that measure the chamber and sample temperatures, respectively. The plant is controlled by a computer that allows the user to input a cooling protocol either manually or as a file prepared in off-line regime. The computer controls the cooling system and forces the chamber temperature to track the prescribed cooling protocol (see Figure 3).

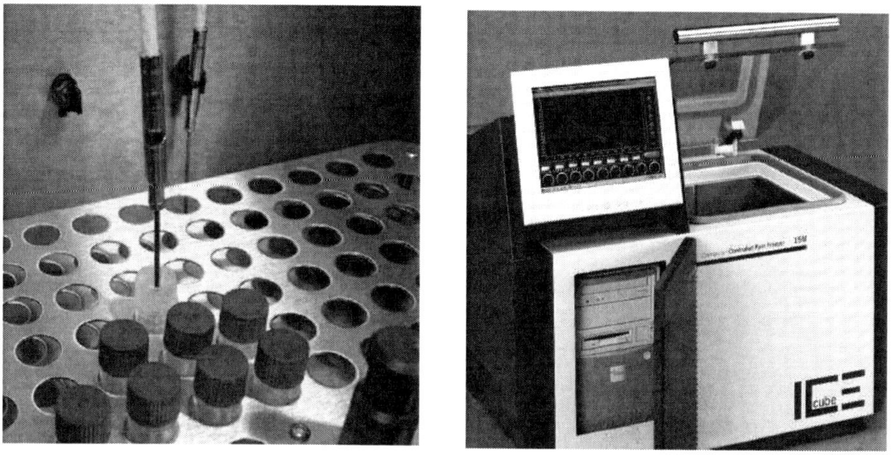

Figure 2. Outlook of the IceCube plant (to the left) and its freezing chamber with two temperature sensors and ampoules containing tissue samples (to the right).

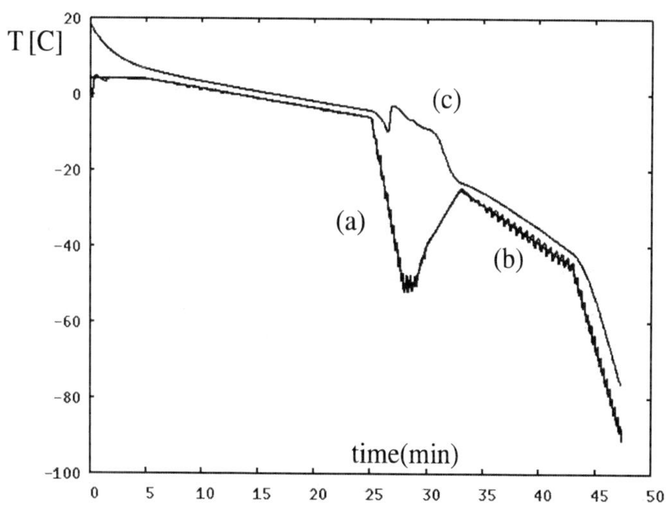

Figure 3. Input and output data of the IceCube plant: (a) A prescribed cooling protocol; (b) The measured chamber temperature (oscillating line) tracking the cooling protocol; (c) The measured sample temperature.

Freezing experiments show an irregular behavior of the temperature near the freezing point because of the release of the latent heat and growth of ice crystals. A typical temperature response of the sample to a constant cooling rate is shown in Figure 4.

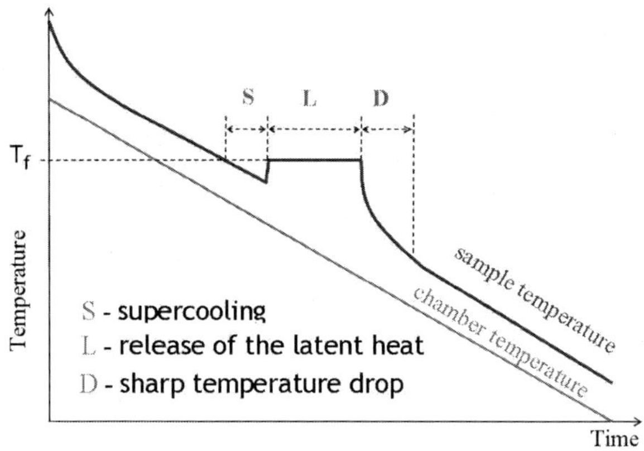

Figure 4. Idealized typical temperature response of the sample when the chamber temperature falls linearly in time. Three dangerous processes are pointed out.

The supercooling process (S) itself is not too dangerous, but in combination with the latent heat release (L) it yields the creation of damaging dendrites: the longer the latent heat release (L) holds, the more dendrites appear. The sudden temperature drop (D) causes a temperature shock to cells. Therefore, it would be preferable to reduce both the duration of the latent heat release (L) and the magnitude of the temperature drop (D).

Consider the following simple model based on averaged thermodynamical values. The heat transfer in the ampoule containing a tissue sample is described by a heat conductivity equation of the form:

$$\frac{\partial H}{\partial t} - \mathbf{div}\,(k\nabla T) = 0,$$

where H and T are the enthalpy density and the temperature, respectively, k is the heat transfer coefficient that may depend on space. The condition on the surface of the ampoule reads:

$$-k\frac{\partial T}{\partial n} = h(T - T_e),$$

where T_e is the chamber temperature, h is the overall heat transfer coefficient. Integration over the volume of the ampoule yields:

$$\frac{d}{dt}\int_V H dV + \int_S h(T - T_e) dS = 0,$$

which can be rewritten as follows:

$$\frac{d}{dt}H = -\alpha(T - T_e). \tag{1}$$

Here, H is now the enthalpy density averaged over the volume of the ampoule, T the boundary temperature averaged over the surface area of the ampoule, $\alpha = |S||V|^{-1}h$, and $|V|$ and $|S|$ denote the volume and the surface area of the ampoule, respectively.

To close equation (1), a constitutive relation $T = T(H)$ is necessary. Such a relation is obtained approximately by recovering the mean surface temperature $T(t)$ of the ampoule from measurements for a given temperature profile $T_e(t)$. Substituting the functions $T(t)$ and $T_e(t)$ in equation (1) and computing $H(t)$ yields the pair $\{T(t), H(t)\}$, which implicitly defines the desired function

$$T = T(H). \tag{2}$$

Note that relation (2) is specific for the sample. Nevertheless such a relation is robust across similar samples. Figure 5 shows the function $T(H)$ obtained for a real ampoule with a milieu containing a tissue sample. Note that the inverse dependence, i.e. the function $H = H(T)$ is not one-valued. That is why the material low (2) is preferable.

The chamber temperature can not change instantaneously. Therefore, equation (1) should be extended by adding the chamber temperature rate u whose maximal absolute value μ is equal about 20°C/s. Thus, we have:

$$\dot{H} = -\alpha(T(H) - T_e), \tag{3}$$

$$\dot{T}_e = u, \quad |u| \leq \mu. \tag{4}$$

The aim of the control u is to smooth the temperature response of the sample on the interval $[t_1, t_2]$ (see Figure 6).

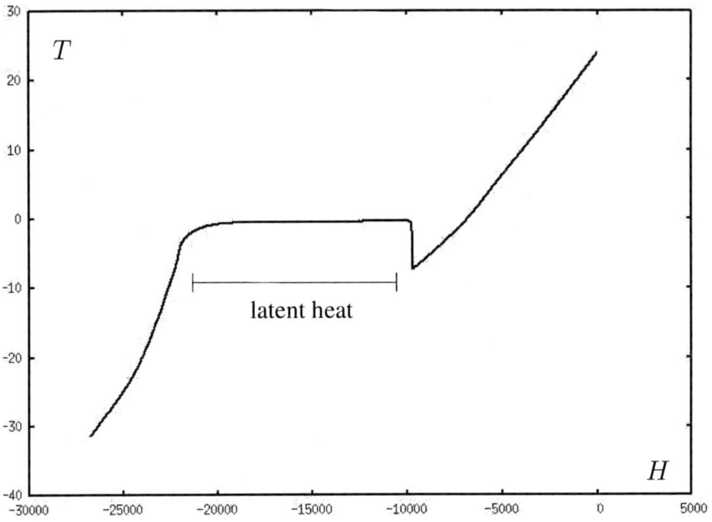

Figure 5. Function $T(H)$ obtained for a real tissue sample.

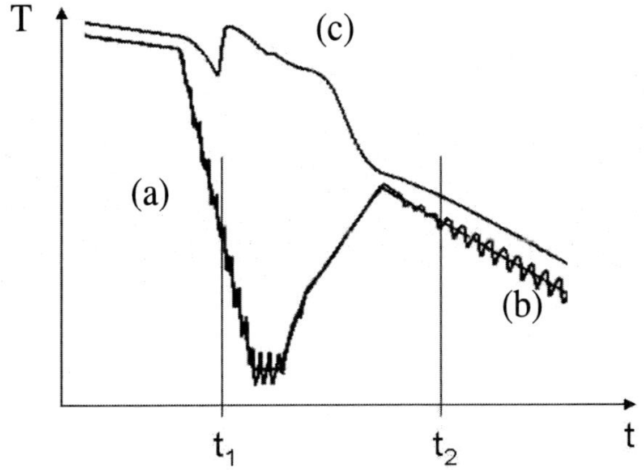

Figure 6. The sample temperature response to a cooling impulse. (a) The cooling impulse; (b) The chamber temperature (oscillating line); (c) The sample temperature response. The idea to smooth the transition to the frozen state using a cooling impulse fails because of wrong parameters of the impulse.

Formally, this is expressed through the minimization of the following performance index:

$$J = \int_{t_1}^{t_2} \left(\frac{d}{dt} T(H(t)) - \theta_0 \right)^2 dt \equiv \int_{t_1}^{t_2} \left(\alpha T_H(H) \cdot (T(H) - T_e) + \theta_0 \right)^2 dt, \quad (5)$$

where θ_0 is a desirable slope of the temperature curve. Ordinary differential equations (3), (4), and functional (5) form a control system, where the control parameter u is the rate of the chamber temperature change. This model possesses the following nice property: the functional J does not depend on the choice of α, whenever H is found from equations (3) and (4), and the constitutive law $T(H)$ is obtained using equation (1). Therefore, the value of α can be simply chosen as $\alpha = 1$, which avoids from the measurement of physical parameters $|V|$, $|S|$, and h.

2.2. Design and Implementation of Optimized Cooling Protocols in a Real Freezing Plant

Let us go back to Section 2.1. Introduce the notation $x = H$, $y = T_e$ and $\phi(x) = T(H)$. Then the system (3) and (4) assumes the form

$$\begin{aligned} \dot{x} &= -\alpha(\phi(x) - y), \\ \dot{y} &= u, \quad |u| \le \mu, \end{aligned} \tag{6}$$

and the functional (5) looks as follows:

$$J = \frac{1}{2} \int_{t_1}^{t_2} (\dot{\phi} - \theta_0)^2 dt, \quad \dot{\phi} = \frac{\partial \phi}{\partial x} \dot{x} = -\alpha \frac{\partial \phi}{\partial x}(x)(\phi(x) - y).$$

The Hamilton function of the control problem (6) is defined as

$$\mathcal{H} = -\alpha p_1 (\phi(x) - y) + p_2 u + \tfrac{1}{2} (\dot{\phi} - \theta_0)^2,$$

and the adjoint equations are given by the relations

$$\begin{aligned} \dot{p}_1 &= -\partial \mathcal{H} / \partial x := \alpha \phi_x p_1 + R_1, \\ \dot{p}_2 &= -\partial \mathcal{H} / \partial y := -\alpha p_1 + R_2, \\ p_1(t_2) &= p_2(t_2) = 0. \end{aligned}$$

Some analysis shows that $\partial\phi/\partial x \geq 0$, $R_1 \geq 0$, $R_2 \leq 0$. This implies that the minimum of J can be achieved using a single cooling impulse (see Figure 7) whose parameters (the start/finish time and the power) can be found through a direct minimization.

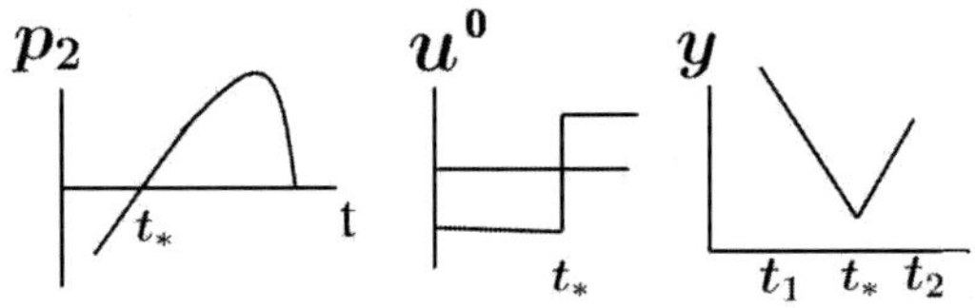

Figure 7. Structure of the optimal control: $u^0(t) \to \max\limits_{|u|\leq\mu} p_2(t)u$. The chamber temperature y shows a cooling impulse.

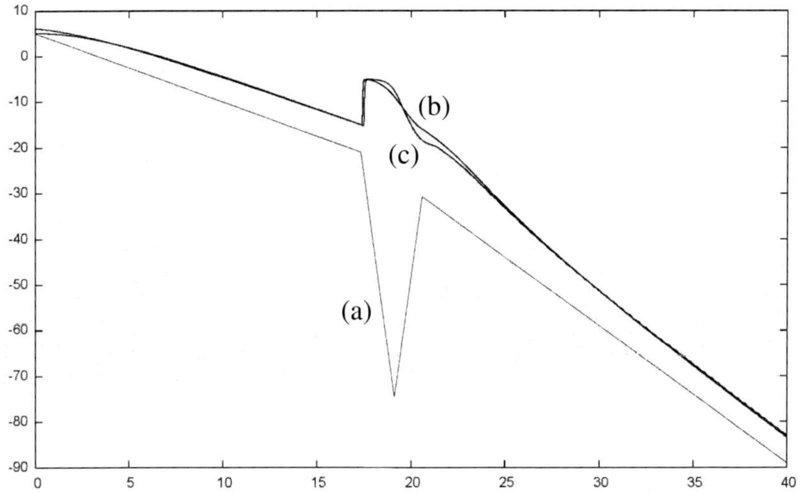

Figure 8. Temperature response when applying an optimized cooling protocol. (a) Optimized cooling impulse; (b) Temperature response measured with the sample sensor, the transition to the frozen state does not show any temperature drop; (c) Computed temperature response.

Figure 8 demonstrates the application of an optimized cooling impulse which provides a good transition regime from the latent heat release to the

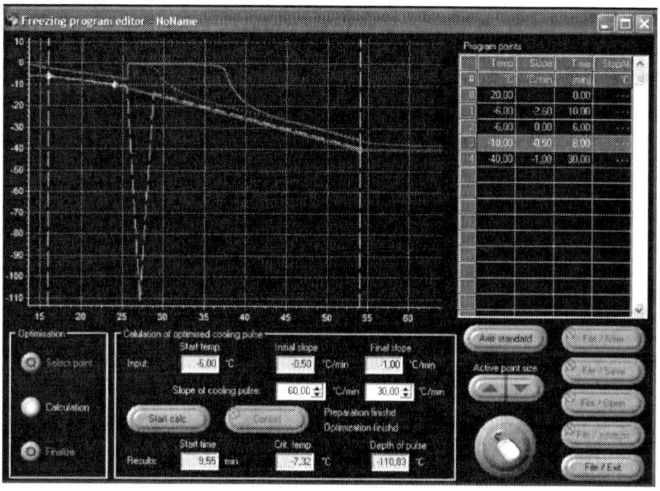

Figure 9. Graphical user interface of an IceCube plant. The temperature response in the case of a constant cooling rate (solid line); optimized cooling protocol with a cooling impulse (dashed line); and the temperature response corresponding to the optimized cooling protocol (dotted line).

frozen state. For comparison, a heuristic (non-optimized) cooling impulse yields a bed result (see Figure 6).

The above sketched optimization techniques are implemented in IceCube freezers. Figure 9 shows the graphical user interface of an IceCube plant. Users can choose an optimization option to compute the optimizing cooling impulse (dashed line). The corresponding optimized temperature response (dotted line) is much better than a non-optimized one (solid line).

2.3. Robustness of Optimal Controls and Hamilton-Jacobi Equations

Assume that the chamber temperature y is measured with an error v so that the control system (6) becomes disturbed as follows:

$$\begin{aligned}
&\dot{x} = -\alpha(\phi(x) - y + v),\\
&\dot{y} = u,\\
&|u| \leq \mu, \quad |v| \leq \nu,
\end{aligned}$$

where ν is a bound on the error. According to the theory of differential games (see e.g. [12]), the control u and the disturbance v are chosen as feedback strategies with the aim to minimize and maximize, respectively, the objective functional. The Hamilton function of the problem is now of the form:

$$\mathcal{H} = -\alpha p_1(\phi(x) - y + v) + p_2 u + \tfrac{1}{2}(\dot{\phi} - \theta_0)^2, \quad \dot{\phi} = \phi_x \dot{x} = -\alpha\phi_x(x)(\phi(x) - y + v).$$

The Hamiltonian is defined by the formula

$$\mathsf{H}(x, y, p_1, p_2) = \max_{|v| \le \nu} \min_{|u| \le \mu} \mathcal{H}(x, y, p_1, p_2, u, v),$$

and the Hamilton-Jacobi (Bellman-Isaacs) equation reads:

$$V_t + \mathsf{H}(x, y, V_x, V_y) = 0, \quad V(t_2, x, y) = 0.$$

It is known that such an equation is generally not solvable in the classical sense. Nevertheless, the equation admits the so-called viscosity solutions (see [13]) that can be numerically computed using special finite-difference grid techniques (see Section 3.1.3). The optimal feedback control is computed using the principle of extremal aiming (see Section 3.1.5).

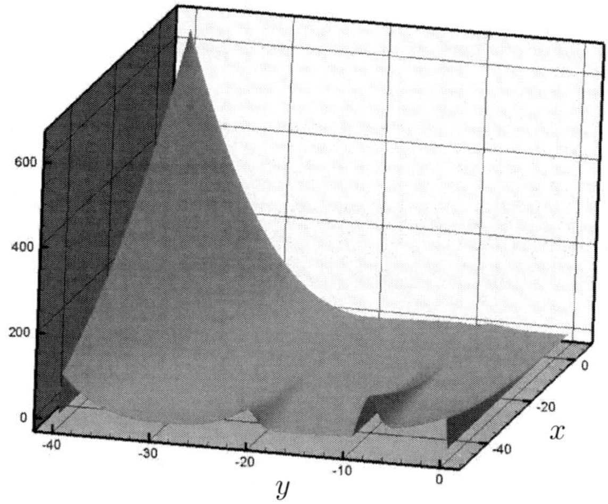

Figure 10. A time snapshot of the value function V.

Figure 10 shows a time snapshot of the approximative value function V. Figure 11 shows simulation results for the optimal feedback controls defined

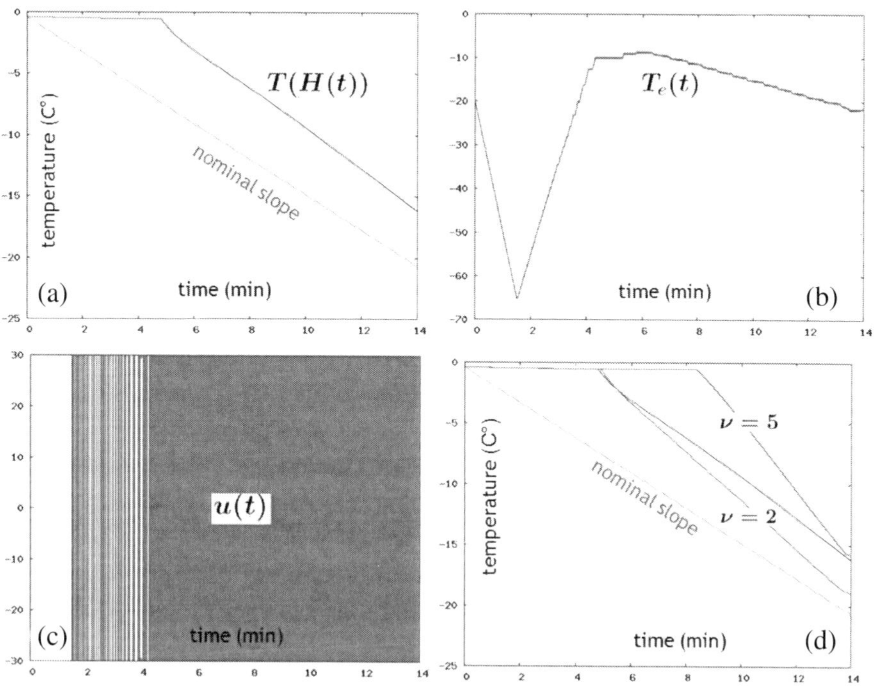

Figure 11. (a) The sample temperature response in the case of optimal feedback control and absence of the disturbance; (b) The chamber temperature profile corresponding to the optimal feedback control; (c) The time progression of the optimal feedback control; (d) The sample temperature responses in the case of optimal feedback control and various values of the disturbance bound ν.

according to Section 3.1.5. Figure 12 shows results in the case of a realistic constitutive relation $\phi(x)$ found from experiments.

2.4. Phase Field Distributed Models of Ice Formation

Nowadays, phase-field models for the simulation of solidification and freezing processes become very popular. They are based on the consideration of the Gibbs free energy which depends on an order parameter that assumes values from -1 (solid) to 1 (liquid) and changes sharply but smoothly over the solidification front so that the sharp liquid-solid interface becomes smoothed. The

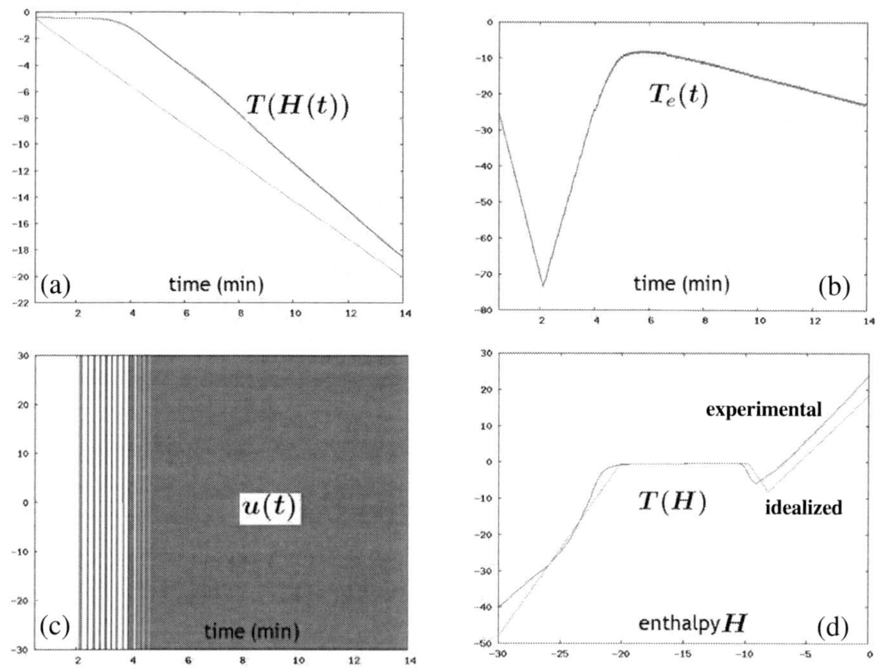

Figure 12. (a) The sample temperature response in the case of optimal feedback control and absence of the disturbance; (b) The chamber temperature profile corresponding to the optimal feedback control; (c) The time progression of the optimal feedback control; (d) Comparison of the idealized function $\phi(x)$ with an experimentally obtained one.

rate of smoothing is controlled by a small parameter, which enables to reach arbitrary approximation of the sharp interface.

We use a phase field model (see [4, 10, 11], and [14]) to describe phase changes in the milieu containing the tissue sample. The control parameter here is the chamber temperature. Notice that the heat flux on the boundary of the ampoule is proportional to the boundary temperature jump. For simplicity, we do not include the solid part (i.e. the tissue sample and the walls of the ampoule) into the model description. Nevertheless, we bear it in mind and account for in numerical simulations.

Let Ω be the interior of the ampoule, Γ the boundary of Ω. The equations look as follows:

$$\vartheta_t + \frac{\ell}{2}\phi_t - K\Delta\vartheta = 0, \qquad x \in \Omega, \tag{7}$$

$$\tau\phi_t - \xi^2\Delta\phi - \frac{1}{2}(\phi - \phi^3) - 2\vartheta = 0, \qquad x \in \Omega, \tag{8}$$

$$-K\frac{\partial\vartheta}{\partial n} = h(\vartheta - \vartheta_e(t) - g), \quad \frac{\partial\phi}{\partial n} = 0, \qquad x \in \Gamma, \tag{9}$$

$$\vartheta|_{t=0} = \vartheta_0 \equiv \text{const} > 0, \quad \phi|_{t=0} = \phi_0 \equiv 1. \tag{10}$$

Here, ϑ is the scaled temperature; ϕ the phase function: $\phi = -1$ for the solid state and $\phi = 1$ for the liquid one; ℓ the scaled latent heat; K the scaled heat conductivity coefficient; h the scaled overall heat conductivity; g the boundary control (variation of the nominal cooling protocol $\vartheta_e(t) = \vartheta_0 + \theta_0 t$, where $\theta_0 < 0$ is a given slope). In contrast to the work [11] we do not assume C^2-regularity of Γ. It is assumed that Γ is of the class $C^{0,1}$, i.e. Lipschitz continuous. This assumption covers various technical designs of ampoules and permits to account for tissue samples of irregular shape.

Proposition 2.1. *If* $\Gamma \in C^{0,1}$, ϑ_0 *and* $\phi_0 \in L^2(\Omega)$, $g \in L^2\big((t_0, t_f); L^2(\Gamma)\big)$, *then exists a unique solution pair* (ϑ, ϕ) *such that*

$$\vartheta,\, \phi \in C(t_0, t_f; L^2(\Omega)) \cap L^2(t_0, t_f; H^1(\Omega)), \quad \phi \in L^4(t_0, t_f; L^4(\Omega)).$$

Here, conventional notations for functional spaces are used (see e.g. [15]). The proof can be found in [16].

Proposition 2.2. *If* $\Gamma \in C^{0,1}$, ϑ_0 *and* $\phi_0 \in H^1(\Omega)$, $g \in L^2\big((t_0, t_f); L^2(\Gamma)\big)$, *then the unique solution pair* (ϑ, ϕ) *satisfies:*

$$\vartheta,\, \phi \in C(t_0, t_f; H^1(\Omega)), \quad \vartheta_t,\, \phi_t \in L^2(0, t_f; L^2(\Omega)), \quad \phi \in L^\infty(t_0, t_f; L^4(\Omega)).$$

Proof. The proof can be done as in [17]. It is only necessary to establish an energy equation for the function ϑ, which is similar to that derived for the function ϕ in [17]. □

Consider first the following functional that estimates the mean quadratic deviation from the nominal cooling protocol $\vartheta_e(t)$:

$$J = \frac{1}{2}\int_{t_0}^{t_f}\int_\Omega (\vartheta - \vartheta_e(t))^2\, dx dt. \tag{11}$$

The adjoint system is derived as in [11] with some modifications related to the boundary control method:

$$\begin{aligned} & -p_t - K\Delta p - 2q = h(\vartheta - \vartheta_e(t)), && x \in \Omega, \\ & -\tau q_t - \frac{\ell}{2}p_t - \xi^2 \Delta q - \frac{1}{2}(1 - 3\phi^2)q = 0, && x \in \Omega, \\ & -K\frac{\partial p}{\partial n} = hp, \quad \frac{\partial q}{\partial n} = 0, && x \in \Gamma, \\ & p_{|t_f} = 0, \quad q_{|t_f} = 0. \end{aligned} \tag{12}$$

Here, p is the adjoint variable corresponding to ϑ; q the adjoint variable related to ϕ.

Proposition 2.3. *If* $\Gamma \in C^{0,1}$, $\vartheta \in L^\infty\big((t_0, t_f); L^2(\Omega)\big)$, $\phi \in L^4\big((t_0, t_f); L^4(\Omega)\big)$, *then there exists a unique solution* (p, q) *such that*

$$p,\, q \in L^\infty(t_0, t_f; L^2(\Omega)) \cap L^2(t_0, t_f; H^1(\Omega)),$$

and therefore $p_{|\Gamma} \in L^2\big((t_0, t_f); L^2(\Gamma)\big)$.

Proof. The proof is similar to the one of proposition 2.1. Note that the term $\phi^2 q$ tested with q yields the positive term $\phi^2 q^2$, which is necessary for the estimate. □

The derivative of the functional J with respect to the control g is defined by the formula

$$J'(\vartheta, \phi)(\delta g) = \int_{t_0}^{t_f} \int_\Gamma \delta g(t, s) p(t, s) ds dt, \quad \text{for all} \quad \delta g \in L^2\big((t_0, t_f) \times \Gamma\big). \tag{13}$$

Therefore, we can identify $J'(\vartheta, \phi)$ with $p|_{(t_0,t_f)\times\Gamma}$ and apply the method of conjugate gradient as follows. Consider the n th step and assume that the control δg^n is already known. Compute the states ϑ^n, ϕ^n, and the adjoint state p^n. Compute δg^{n+1} as

$$\delta g^{n+1} = \delta g^n + \alpha^n d^n,$$

where α^n is found as an approximate solution of the line search problem $\alpha^n \to \min_\alpha J(g^n + \alpha d^n)$, and the conjugate direction d^n is found from the relations (see [18])

$$d^n = -p^n + \beta^n d^{n-1}, \quad \beta^n = \left[\int_{t_0}^{t_f} \int_\Gamma (p^{n-1})^2 ds dt\right]^{-1} \int_{t_0}^{t_f} \int_\Gamma (p^n)^2 ds dt.$$

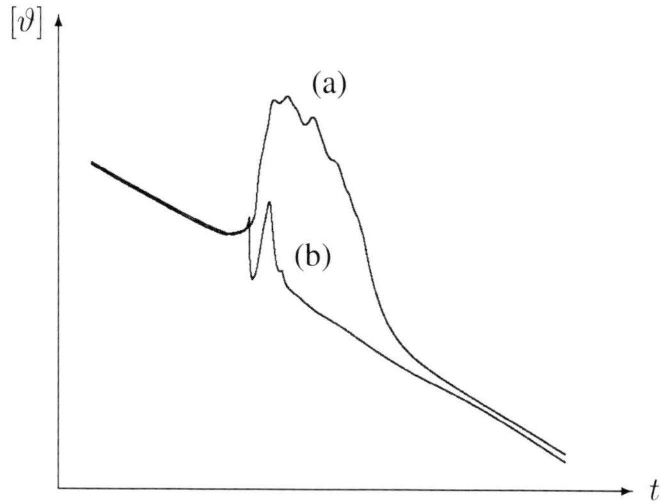

Figure 13. The mean temperature response, $[\vartheta]$. (a) Without control; (b) With control obtained from the minimization of functional (11).

Figure 13 shows the result of minimization of functional (11).

It should be noticed that curve (b) shows oscillations about the nominal protocol $\vartheta_e(t)$. Unfortunately, these oscillations are too sharp so that the control obtained is unusable. To obtain more appropriate results, a modification of the functional (11) is necessary.

Denote $\mathbb{H} = L^2(\Omega)$, $\mathbb{V} = H^1(\Omega)$, for brevity. Let $(\cdot, \cdot)$ be the inner product of $L^2(\Omega)$, $\Lambda : \mathbb{V} \to \mathbb{V}'$ the canonical isomorphism so that $\|f\|^2_{\mathbb{V}'} = \langle \Lambda^{-1} f, f \rangle$, where the angle brackets denote the duality between $\mathbb{V}$ and $\mathbb{V}'$. Consider the following functional that measures a mean quadratic deviation of the time derivative of the solution from the given slope θ_0:

$$J = \frac{1}{2} \int_0^{t_f} \|\vartheta_t - \theta_0\|^2_{\mathbb{V}'} dx dt. \tag{14}$$

The formal derivation of the adjoint equations for this functional yields:

$$
\begin{aligned}
&-p_t - K\Delta p - 2q = -h\big[\Lambda^{-1}(\vartheta_t - \theta_0)\big]_t,\\
&-\tau q_t - \frac{\ell}{2}p_t - \xi^2\Delta q - \frac{1}{2}(1-3\phi^2)q = 0,\\
&-K\frac{\partial p}{\partial n} = hp, \quad \frac{\partial q}{\partial n} = 0,\\
&\quad p(t_f) = h\Lambda^{-1}(\vartheta_t - \theta_0)|_{t_f}, \quad \tau q(t_f) + \frac{\ell}{2}p(t_f) = 0.
\end{aligned}
\tag{15}
$$

This problem does not have conventional solutions in the sense of propositions 2.1 and 2.2 because the right-hand side of the first equation does not possess sufficient regularity. To formulate this problem correctly, use the method of transposed isomorphism developed in [15] and [19]. Define

$$
W_0(0,t_f) = \left\{\eta : \eta \in L^2\left(0,t_f;\mathbb{V}\right),\ \eta_t, \Delta\eta \in L^2\left(0,t_f;\mathbb{V}'\right), -K\frac{\partial\eta}{\partial n} = h\eta,\ \eta(0) = 0\right\},
$$

$$
Z_0(0,t_f) = \left\{\zeta : \zeta \in L^2\left(0,t_f;\mathbb{V}\right),\ \zeta_t, \Delta\zeta \in L^2\left(0,t_f;\mathbb{V}'\right),\ \frac{\partial\zeta}{\partial n} = 0,\ \zeta(0) = 0\right\}.
$$

The adjoint system is obtained by multiplying the equations of system (15) by test functions and double partial integration:

$$
\begin{aligned}
&\int_0^{t_f} \left[\langle p, \eta_t - K\Delta\eta\rangle - 2(q,\eta)\right] dt = \int_0^{t_f} \langle \Lambda^{-1}(\vartheta_t - \theta_0), \eta_t\rangle dt,\\
&\int_0^{t_f} \left[\langle q, \tau\zeta_t - \xi^2\Delta\zeta\rangle + \frac{\ell}{2}\langle p, \zeta_t\rangle - \frac{1}{2}\left((1-3\phi^2)q, \zeta\right)\right] dt = 0.
\end{aligned}
\tag{16}
$$

Here, η and ζ are the test functions from the spaces $W_0(0,t_f)$ and $Z_0(0,t_f)$, respectively. The unique solvability is stated using the fact that the mappings

$$
\ell^{-1} : \eta \to \eta_t - K\Delta\eta : \quad W_0(0,t_f) \to L^2\left(0,t_f;\mathbb{V}'\right)
$$

$$
\wp^{-1} : \zeta \to \tau\zeta_t - \xi^2\Delta\zeta : \quad Z_0(0,t_f) \to L^2\left(0,t_f;\mathbb{V}'\right).
$$

are isomorphisms.

Proposition 2.4. *If $\vartheta_t \in L^2\left(0,t_f;\mathbb{V}'\right)$, then system (16) has a unique solution pair (p,q) such that $p,q \in L^2\left(0,t_f;\mathbb{V}\right)$, and therefore $p_{|\Gamma} \in L^2\big((0,T);L^2(\Gamma)\big)$. The derivative of functional (14) is given by (13).*

The proof of this proposition can be found in [20].

Consider now the following functional that seems to be more natural than (14):

$$J = \frac{1}{2}\int_0^{t_f} \|\vartheta_t - \theta_0\|_{\mathbb{H}}^2 dt. \tag{17}$$

To formulate the adjoint equations, we need now the following spaces:

$$W_0(0,t_f) = \left\{\eta : \eta \in L^\infty(0,t_f;\mathbb{V}),\ \eta_t, \Delta\eta \in L^2(0,t_f;\mathbb{H}), -K\frac{\partial\eta}{\partial n} = h\eta,\ \eta(0) = 0\right\},$$

$$Z_0(0,t_f) = \left\{\zeta : \zeta \in L^\infty(0,t_f;\mathbb{V}),\ \zeta_t, \Delta\zeta \in L^2(0,t_f;\mathbb{H}),\ \frac{\partial\zeta}{\partial n} = 0,\ \zeta(0) = 0\right\}.$$

The adjoint equations are:

$$\begin{aligned}
&\int_0^{t_f} [(p, \eta_t - K\Delta\eta) - 2(q,\eta)]\, dt = \int_0^{t_f} (\vartheta_t - \theta_0, \eta_t) dt, \\
&\int_0^{t_f} \left[(q, \tau\zeta_t - \xi^2\Delta\zeta) + \frac{\ell}{2}(p,\zeta_t) - \frac{1}{2}\left((1-3\phi^2)q, \zeta\right)\right] dt = 0.
\end{aligned} \tag{18}$$

The following result can be proved analogously to proposition 2.4.

Proposition 2.5. *If* $\vartheta_t \in L^2(0,t_f;\mathbb{H})$, *then system (18) has a unique solution pair* (p,q) *such that* $p, q \in L^2(0,t_f;\mathbb{H})$.

The trouble here is that the derivative of the functional (17) should be defined by (13), but the trace p_Γ can not be defined properly because of insufficient regularity of p. To overcome this difficulty, consider the following functional that estimates the deviation of the time derivative of the mean temperature from the nominal slope:

$$J = \frac{1}{2}\int_0^{t_f} ([\vartheta]_t - \theta_0)^2 dt, \quad \text{where} \quad [\vartheta] = \frac{1}{|\Omega|}\int_\Omega \vartheta dx. \tag{19}$$

Now, express $[\vartheta]_t$ through values that do not contain time derivatives. To do that, integrate the model equations (7) and (8) over Ω to get:

$$\begin{aligned}
&[\vartheta]_t - \theta_0 = \\
&-\frac{1}{|\Omega|}\left[h\int_\Gamma [\vartheta - \vartheta_e(t) - g]ds + \frac{1}{4\tau}\int_\Omega (\phi - \phi^3)dx + \frac{1}{\tau}\int_\Omega \vartheta dx\right] - \theta_0 =: \gamma_{\vartheta,\phi,g}(t).
\end{aligned} \tag{20}$$

The adjoint system and the derivative of the functional are given by:

$$
\begin{aligned}
& -p_t - K\Delta p - 2q = \frac{1}{\tau}\gamma_{\vartheta,\phi,g}(t), && x \in \Omega, \\
& -\tau q_t - \frac{\ell}{2}p_t - \xi^2\Delta q - \frac{1}{2}(1-3\phi^2)q = \frac{1}{4\tau}\gamma_{\vartheta,\phi,g}(t)(1-3\phi^2), && x \in \Omega, \\
& -K\frac{\partial p}{\partial n} = hp + h\gamma_{\vartheta,\phi,g}(t), \quad \frac{\partial q}{\partial n} = 0, && x \in \Gamma, \\
& \quad p(t_2) = 0, \quad q(t_2) = 0,
\end{aligned}
\tag{21}
$$

$$
J'(\vartheta,\phi,g)(\delta g) = \frac{h}{|\Omega|}\int_{t_0}^{t_f}\int_\Gamma \delta g(t,s)[p(t,s) - \gamma_{\vartheta,\phi,g}(t)]dsdt, \qquad \forall\, \delta g \in L^2\big((t_0,t_f)\times\Gamma\big). \tag{22}
$$

System (21) is well posed because $\gamma_{\vartheta,\phi,g}(t) \in L^2(0,t_f)$ and $\phi \in L^\infty(0,t_f;L^4(\Omega))$ (see Proposition 2.2). Thus, the same regularity of p as in Proposition 2.3 holds, and therefore formula (22) is well defined. The result of the application of such a technique is shown in Figure 14.

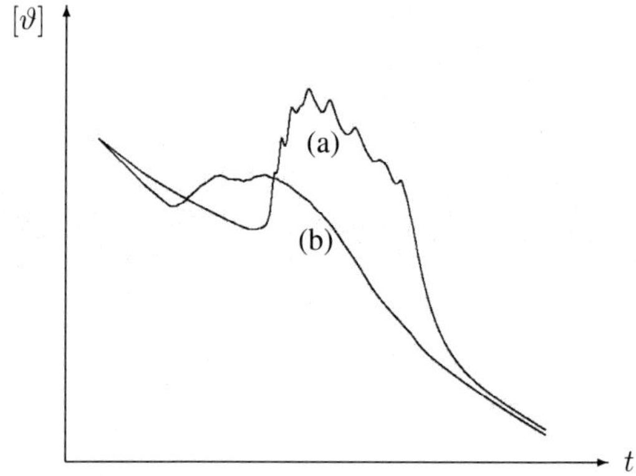

Figure 14. The mean temperature response, $[\vartheta]$. (a) Without control; (b) With control obtained from the minimization of functional (19).

Notice, that all optimal controls previously discussed in this subsection are functions of space on the boundary, which can hardly be implemented technically. Therefore, it would be reasonable to look for a control which is constant

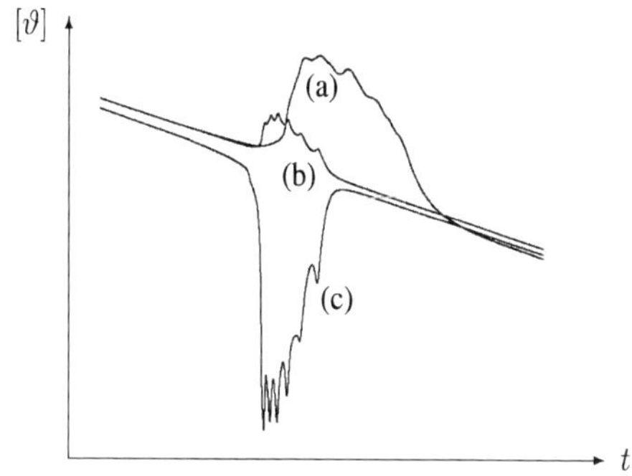

Figure 15. The mean temperature response, $[\vartheta]$. (a) Without control; (b) With the feedback control (23); (c) Control protocol.

on the boundary so that g is a function of t only. Compute such a control from the condition $\gamma_{\vartheta,\phi,g}(t) = 0$, i.e. $[\vartheta]_t - \theta_0 = 0$ (see relation (20)). This yields the following feedback rule:

$$g(t) = \frac{1}{h|\Omega||\Gamma|}\left[h\int_\Gamma [\vartheta - \vartheta_e(t)]ds + \frac{1}{4\tau}\int_\Omega (\phi - \phi^3)dx + \frac{1}{\tau}\int_\Omega \vartheta dx\right] + \frac{1}{h|\Gamma|}\theta_0. \tag{23}$$

Figure 15 demonstrates the application of heuristic rule (23). The result seems to be comparable with that obtained using techniques based on the minimization of functional (19), but the runtime is much shorter.

2.5. Ice Formation in the Extracellular Matrix

The description of ice formation in the extracellular matrix of a tissue is based on the thermodynamics of porous media (see [21]). The main feature of porous or soil media is that the unfrozen water content is a function of the temperature only. Such a function can be considered as a material property that depends on the pore size distribution and the material of the solid matrix. The interaction of the liquid with the solid matrix causes homogeneous nucleation of the liquid and, therefore, excludes supercooling. In freezing soil science, the unfrozen water content is measured directly by NMR (Nuclear Magnetic Resonance). There are a lot of works devoted to the theoretical derivation of the unfrozen water content function (see e.g. [21] and [22]).

Introduce some notation. Let T be the absolute temperature, β_ℓ the liquid water volume fraction (unfrozen water content), β_i the ice volume fraction, β_m the volume fraction of the solid matrix, ρ the density (assume that the densities of the liquid and ice phases are equal), L the specific phase change latent heat, C the specific heat capacity (assume that the specific heat capacities of the liquid and ice phases are equal), T_0 the freezing temperature. Obviously, $\beta_\ell + \beta_i = 1 - \beta_m$. Assuming that $\beta_m = const$ and scaling the last equation yields: $\beta_\ell + \beta_i = 1$.

According to [21], the density of the free energy of the liquid and ice phases are given by:

$$\begin{aligned} \Psi_\ell &= \rho\beta_\ell \left\{ -\frac{L}{T_0}(T - T_0) + T h_\ell(\beta_\ell, \beta_i) - CT \log T \right\}, \\ \Psi_i &= \rho\beta_i \left\{ T h_i(\beta_\ell, \beta_i) - CT \log T \right\}, \end{aligned} \tag{24}$$

where $h_\ell(\beta_\ell, \beta_i)$ and $h_i(\beta_\ell, \beta_i)$ are functions that account for the interaction of the liquid and ice phases, respectively, with the solid matrix. The total free energy density is given by the formula

$$\Psi = \rho\beta_\ell \left\{ -\frac{L}{T_0}(T - T_0) \right\} + \rho T \left\{ \beta_\ell h_\ell(\beta_\ell, \beta_i) + \beta_i h_i(\beta_\ell, \beta_i) \right\} - \rho CT \log T.$$

The second law of thermodynamics yields the model equation:

$$\frac{de}{dt} = -\mathbf{div} q + r, \quad q = -\mathcal{K}\nabla T, \quad \text{with} \quad e = \Psi - T\frac{\partial \Psi}{\partial T}. \tag{25}$$

To determine the function defining the unfrozen water content, assume (see chapter 11.3 of [21]) that the phase transition itself is not dissipative, and the dissipation is caused by thermal conditions only. This results in the equilibrium of the driving forces of the phase change:

$$\frac{\partial \Psi_\ell}{\partial \beta_\ell} = \frac{\partial \Psi_i}{\partial \beta_i} \quad \text{if} \quad 0 < \beta_\ell < 1,$$

which is equivalent to the relation

$$\frac{L}{T_0}(T - T_0) = T\frac{\partial \tilde{h}(\beta_\ell)}{\partial \beta_\ell}, \quad \text{if} \quad 0 < \beta_\ell < 1, \tag{26}$$

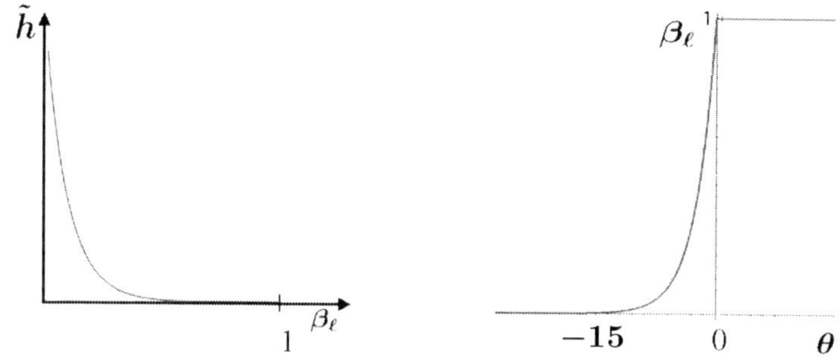

Figure 16. Functions defining the unfrozen water content in a porous medium.

where

$$\tilde{h}(\beta_\ell) = \beta_\ell h_\ell(\beta_\ell, \beta_i) + \beta_i h_i(\beta_\ell, \beta_i)|_{\beta_i = 1-\beta_\ell}.$$

Using (25), (26), and introducing the Celsius temperature $\theta := T - T_0$ yields the following mathematical model:

$$\begin{aligned} \rho C \frac{\partial \theta}{\partial t} + \rho L \frac{\partial \beta_\ell}{\partial t} - \mathcal{K}\Delta\theta = 0, \\ -\mathcal{K}\frac{\partial \theta}{\partial n} = \lambda(\theta - \theta_0 t - g), \end{aligned} \tag{27}$$

$$\beta_\ell = \phi\left(\frac{L\theta}{T_0(T_0+\theta)}\right), \quad \text{with} \quad \phi(y) = \left[\frac{\partial \tilde{h}(\beta_\ell)}{\partial \beta_\ell}\right]^{-1}(y). \tag{28}$$

Here, g is the boundary control (variation of the nominal cooling protocol $\theta_0 t$, where $\theta_0 < 0$ is a given slope). The function β_ℓ (see Figure 16) was recovered from data obtained in experiments with tissue samples on an IceCube plant (see section 2.1). Figure 17 shows the mean temperature response when $g \equiv 0$. Figure 18 shows a time snapshot of the spatial distribution of the temperature and unfrozen water content in this case.

Note that supercooling can not occur in the model (27) and (28). Therefore, it would be reasonable to use the following simple functional to smooth the transition from the latent heat release to the frozen state:

$$J = \frac{1}{2}\int_0^{t_f}\int_\Omega (\theta - \theta_0 t)^2 dx\, dt. \tag{29}$$

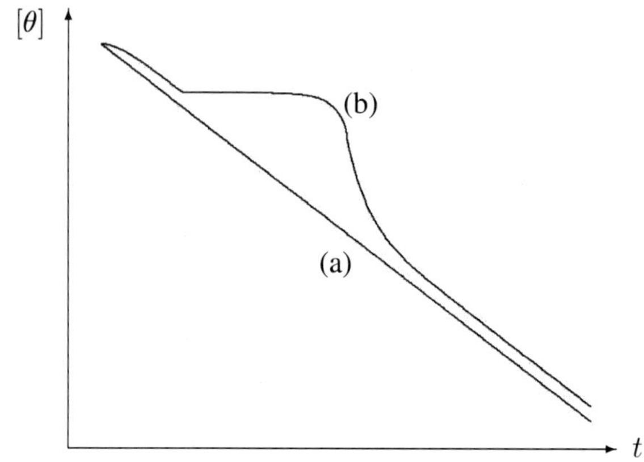

Figure 17. (a) The nominal cooling protocol; (b) The mean temperature response.

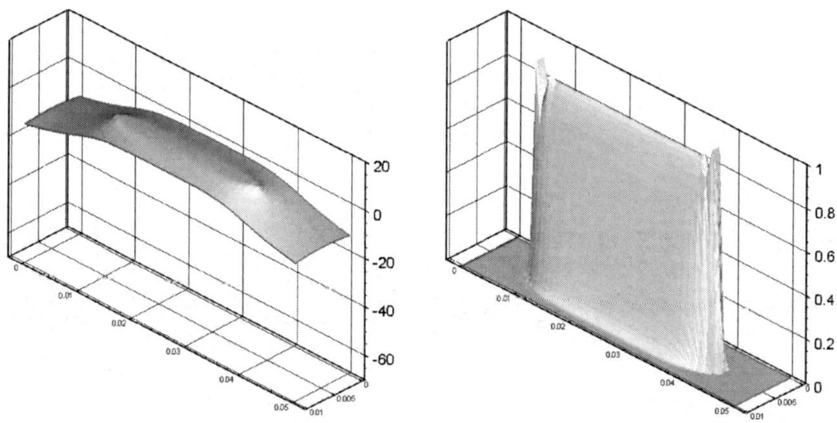

Figure 18. A time snapshot of the spatial distribution of the temperature (left) and the unfrozen water content (right) in the case of the nominal cooling protocol.

The adjoint equation looks as follows:

$$
\begin{aligned}
&-\left[\rho C + \phi_y\left(\frac{L\theta}{T_0(\theta+T_0)}\right)\right]\frac{\partial p}{\partial t} - \mathcal{K}\Delta p = \lambda\big(\theta - \theta_0 t\big), && x \in \Omega,\\
&-\mathcal{K}\frac{\partial p}{\partial n} = \lambda p, && x \in \partial\Omega,\\
&\; p(t_f) = 0.
\end{aligned}
\tag{30}
$$

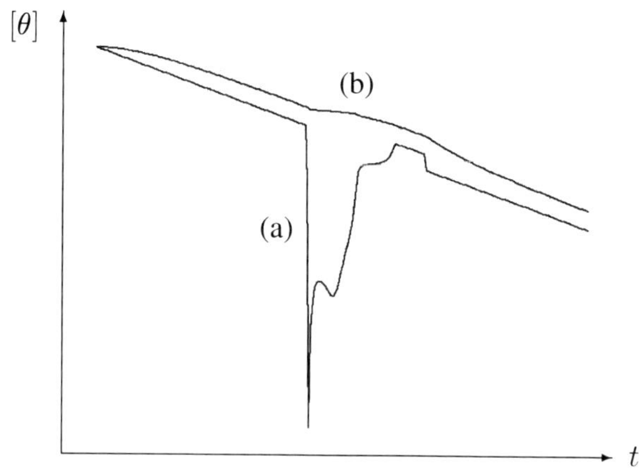

Figure 19. (a) Optimal cooling protocol obtained from the minimization of functional (29); (b) The mean temperature response.

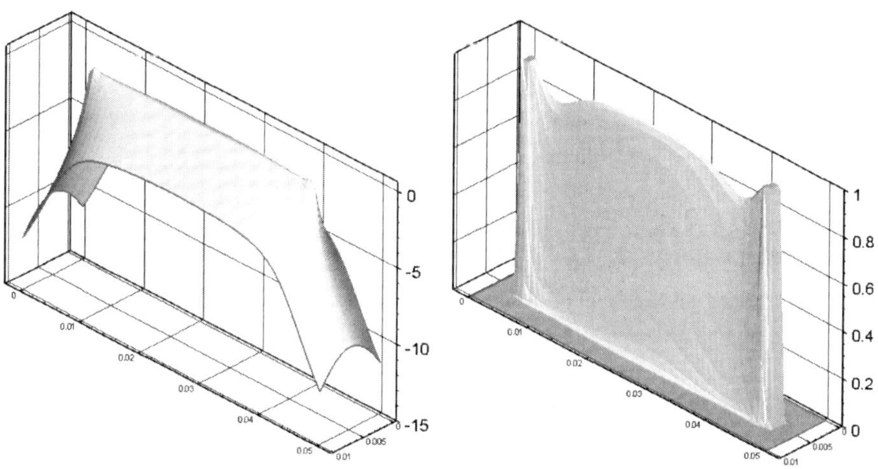

Figure 20. A time snapshot of the spatial distribution of the temperature (left) and the unfrozen water content (right) in the case of optimal cooling protocol obtained from the minimization of functional (29).

The derivative of the functional is given (up to notations) by (13). Figure 19 shows the mean temperature response in the case of optimal control obtained from the minimization of functional (29). Figure 20 shows a time snapshot of the spatial distribution of the temperature and unfrozen water content in this case (compare with Figure 18).

2.6. Stresses in a Frozen Porous Medium

It is well known that the increase of the volume during the water to ice phase change is rather large. If the phase change occurs in a pore of a porous material, a very large stress can be exerted on the pore walls (see Figure 21).

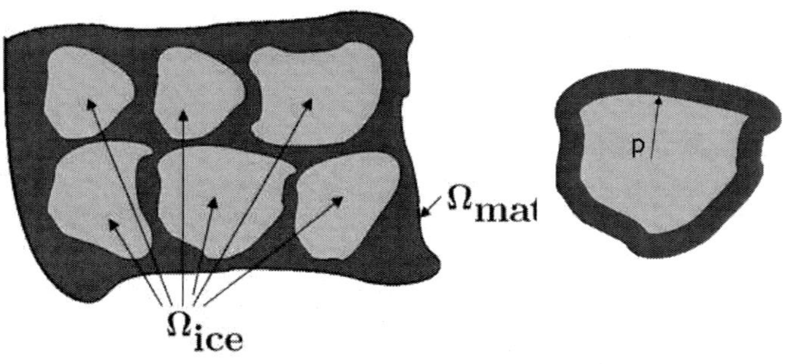

Figure 21. Porous medium filled with a liquid (left); pressure in the pore (right).

Let Ω_{ice} and Ω_{mat} be domains occupied by the ice-water mixture and the solid matrix, respectively. It is easy to show that the displacement vector $\vec{u}$ is governed by the following equation given in a weak formulation:

$$\int_{\Omega_{\text{ice}}} C^{\text{ice}}_{ijkl}\tilde{\epsilon}_{ij}(\vec{u},\beta_\ell)\epsilon_{kl}(\vec{\varphi}) + \int_{\Omega_{\text{mat}}} C^{\text{mat}}_{ijkl}\epsilon_{ij}(\vec{u})\epsilon_{kl}(\vec{\varphi}) = 0. \tag{31}$$

Here, C^{ice}_{ijkl} and C^{mat}_{ijkl} are the elastic stiffness tensors of ice and the solid matrix material, respectively, $\vec{\varphi}(x)$ the test vector function,

$$\epsilon_{ij}(\vec{u}) = \frac{1}{2}\left(\partial u_i/\partial x_j + \partial u_j/\partial x_i\right), \; \tilde{\epsilon}_{ij}(\vec{u},\beta_\ell) = \epsilon_{ij}(\vec{u}) - \epsilon^{tr}_{ij}(\beta_\ell), \; \epsilon^{tr}_{ij}(\beta_\ell) = \frac{1}{3}\alpha(1-\beta_\ell)\delta_{ij},$$

and α the expansion coefficient (the relative volume increase during the phase change). Obviously, these equations are not appropriate for numerical implementations because of a fine structure of the domains Ω_{ice} and Ω_{mat}. To state a computable model, homogenization techniques (see [23, 24, 25], and [26]) are to be used. A porous medium is a probabilistic composite material, which demands the use of stochastic homogenization techniques (see e.g. [25]). Nevertheless, the assumption of periodicity provides a reasonable first approximation. Suppose the structural cell of the porous material can be scaled to the unit cube $Y = [0,1]^3$. Let $Y_{\text{ice}} \subset Y$ be the part occupied by the ice-water mixture, $Y_{\text{mat}} = Y \setminus Y_{\text{ice}}$ the part occupied by the solid matrix, $I_{Y_{\text{mat}}}(y)$ and $I_{Y_{\text{ice}}}(y)$ are Y-periodically continued indicator functions of Y_{mat} and Y_{ice}, respectively (see Figure 22).

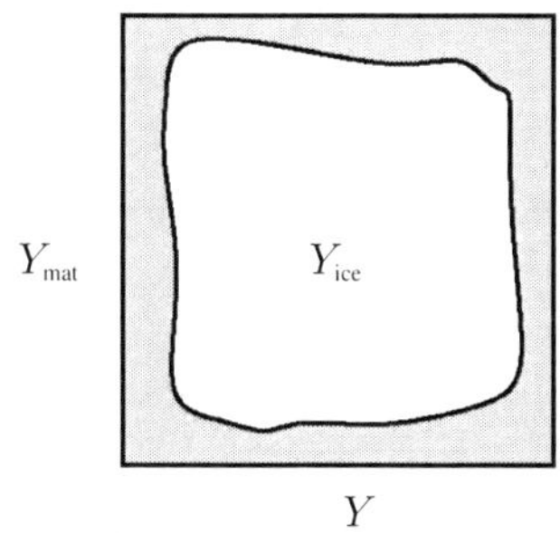

Figure 22. Scaled structural cell Y.

Let ε be the parameter of homogenization. Then the system (31) can be rewritten as follows

$$\int_\Omega \Big[C_{ijkl}(x/\varepsilon)\epsilon_{ij}(\vec{u}) - \alpha(1-\beta_\ell)I_{Y_{\text{ice}}}(x/\varepsilon)\delta_{ij}\Big]\epsilon_{kl}(\vec{\varphi}) = 0, \qquad (32)$$

where

$$C_{ijkl}(y) = I_{Y_{\text{ice}}}(y)C^{\text{ice}}_{ijkl} + I_{Y_{\text{mat}}}(y)C^{\text{mat}}_{ijkl}.$$

Using techniques of [26], it is easy to show that the limiting equation, as $\varepsilon \to 0$, reads:

$$\int_\Omega \Big[D_{ijkl}\,\epsilon_{ij}(\vec{u}) - \alpha(1-\beta_\ell)d_{kl}\Big]\epsilon_{kl}(\vec{\varphi}) = 0, \qquad (33)$$

where the constant tensors D and d are given by the formulae

$$\begin{aligned} D_{klmn} &= \int_Y C_{ijkl}(y)\Big[\delta_{im}\delta_{jn} + \epsilon_{ij}(\vec{N}_{mn}(y))\Big]\,dy, \\ d_{kl} &= \int_Y C_{ijkl}(y)\Big[\epsilon_{ij}(\vec{M}(y)) - \delta_{ij}I_{Y_{\text{ice}}}(y)\Big]\,dy. \end{aligned} \qquad (34)$$

Here, the auxiliary functions $\vec{N}_{mn}(y)$ and $\vec{M}(y)$ satisfy the following cell equations:

$$\int_Y C_{ijkl}(y)\left[\delta_{im}\delta_{jn}+\epsilon_{ij}(\vec{N}_{mn}(y))\right]\epsilon_{kl}\big(\vec{\psi}(y)\big)dy=0,\quad n=1,2,3,\ m\le n,$$
$$\int_Y C_{ijkl}(y)\left[\epsilon_{ij}(\vec{M}(y))-\delta_{ij}I_{Y_{\text{ice}}}(y)\right]\epsilon_{kl}\big(\vec{\psi}(y)\big)dy=0, \tag{35}$$

where $\vec{N}_{mn}(y)=\vec{N}_{nm}(y)$, if $m>n$, $\vec{N}_{mn}$ and $\vec{M}$ are Y-periodic, and $\vec{\psi}(y)$ are Y-periodic test functions.

For simplicity, coupled equations (33) and (27) were numerically implemented in the case of two dimensions. Figure 23 shows a time snapshot of the stress distribution in Ω.

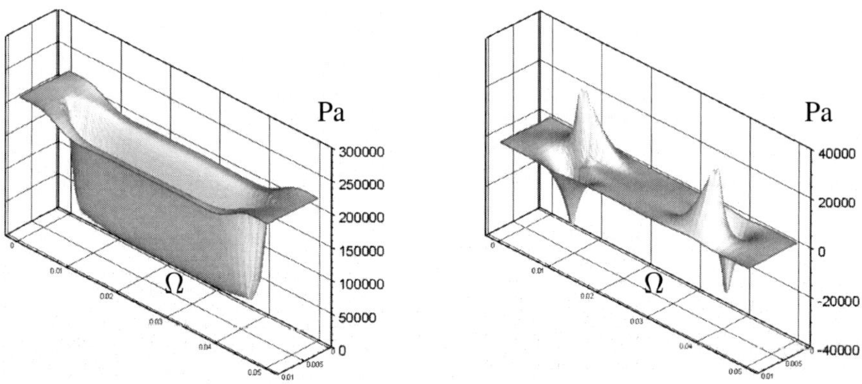

Figure 23. A time snapshot of the stresses in a porous medium under freezing: the pressure (left); the shear stress (right).

Consider now stresses exerted on a sole cell located in a pore with freezing water (see Figure 24).

The cell membrane is a complicated structure that inhibits the heat flux. Therefore, the liquids phases inside and outside the cell freeze not simultaneously, which results in the development of high stresses. The displacement vec-

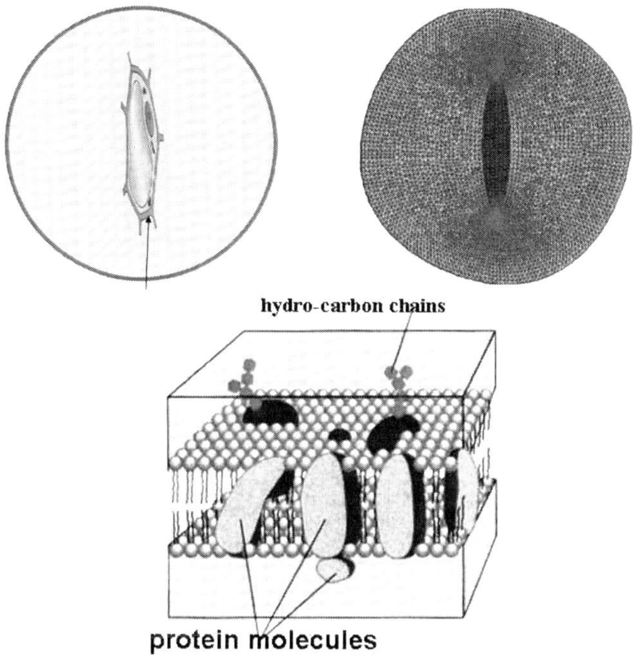

Figure 24. Pore with a cell inside (at the upper left), finite element model (at the upper right), structure of the cell membrane (below).

tor is described by the following equation:

$$\int_{\Omega_{\text{ice}}} C^{\text{ice}}_{ijkl}\tilde{\epsilon}_{ij}(\vec{u},\beta_\ell)\epsilon_{kl}(\vec{\varphi}) + \int_{\Omega_{\text{cell}}} C^{\text{cell}}_{ijkl}\epsilon_{ij}(\vec{u})\epsilon_{kl}(\vec{\varphi}) = 0, \tag{36}$$

where C^{cell}_{ijkl} is the stiffness tensor describing averaged material of the cell. Properties of this tensor are similar to these established in [27] for a limiting tensor obtained when averaging a liquid-solid structure under oscillation. The main feature of such a tensor is the absence of shear stiffness. Two dimensional simulations of equation (36) coupled with equation (27) are shown on Figures 25 and 26. The first figure shows the temperature and phase jumps on the cell boundary. The second one represents the jump of the pressure.

Figure 27 explains the idea of balancing the pressures on the cell bound-

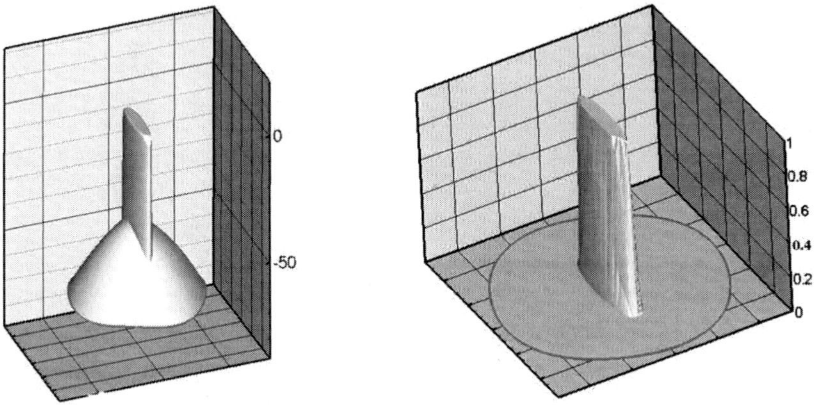

Figure 25. A time snapshot of the temperature (left) and the unfrozen water content (right) in the cell.

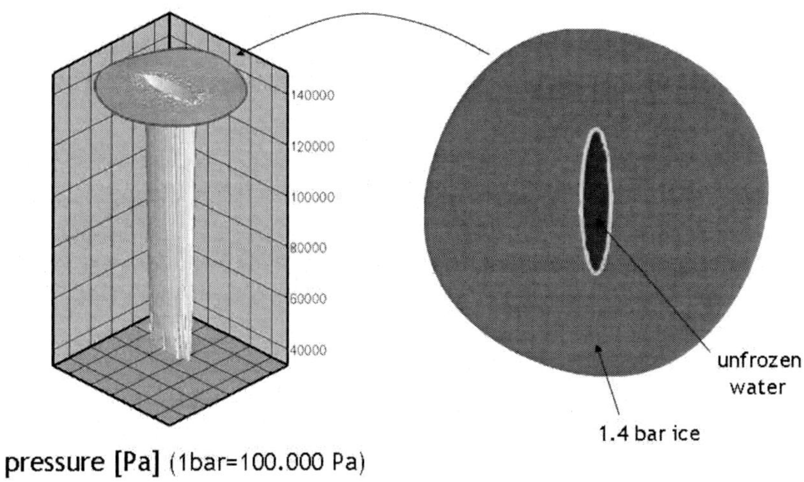

Figure 26. Pressure exerted on the cell boundary.

ary: liquids inside and outside the cell must freeze simultaneously. This can be achieved through lowering the freezing point of the outside liquid, i.e. the relation $T_0^{\text{ins}} > T_0^{\text{out}}$ should hold. An appropriate choice of the control parameter $\delta T_0 = T_0^{\text{ins}} - T_0^{\text{out}}$ provides a good result.

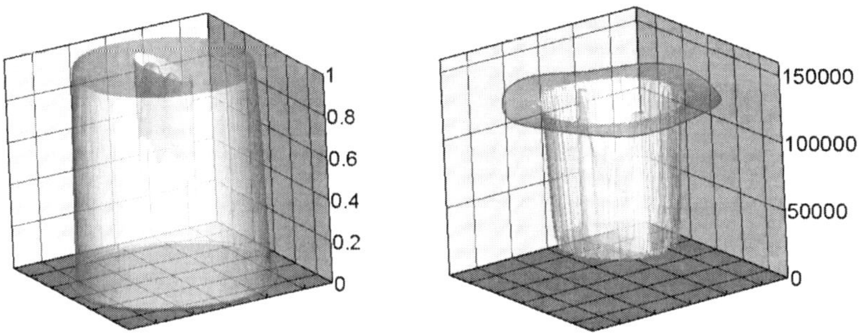

Figure 27. Balancing the pressures on the cell boundary through lowering the freezing point of the extracellular liquid. Time snapshots of the jump in the unfrozen liquid content (left) and in the pressure (right) on the cell boundary are shown.

3. Consideration of Damaging Processes on Cellular Level

This section considers mathematical models describing the processes of freezing and thawing inside of pores of the extracellular matrix. Two injuring factors are considered. One of them is a large stress exerted on cell membranes. Another injuring factor is excessive shrinkage and swelling due to osmotic dehydration and rehydration occurring during the freezing and warming phases, respectively.

Large stresses exerted on cell membranes occur at slow cooling because of non-simultaneous freezing of extracellular and intracellular fluids. The use of rapid cooling rates is not a perfect solution because intracellular water forms small, irregularly-shaped ice crystals (dendrites) that are relatively unstable. If frozen cells are subsequently thawed, dendrites will aggregate to form larger, more stable crystals that may cause damage. Therefore, it is reasonable to apply control theory to provide simultaneous freezing of extracellular and intracellular fluids even for slow cooling rates. Thus, mathematical models containing control variables and optimization criteria are to be formulated.

Note that models of phase transitions basically utilize partial differential equations that describe the dynamics of phases in each spatial point (see e.g. [10], [21], and [11]). Nevertheless, the spatial distribution is not very important, if small objects such as living cells and pores of the extracellular matrix are investigated. Our experience shows that appropriate averaging techniques accurately reduce spatially distributed models to a few ordinary differential equations with control parameters and uncertainties. However, such equations contain, as a rule, nonlinear dependencies given by tabular data. This complicates the application of traditional control design methods based on Pontryagin's maximum principle. Nevertheless, dynamic programming methods related to Hamilton-Jacobi-Bellman-Isaacs (HJBI) equations are suitable, if stable grid algorithms for solving HJBI problems are available. This section describes stable grid procedures that enable to design optimized controls (cooling protocols) for ODE systems describing competitive ice formation inside and outside of living cells.

Shrinkage and swelling of cells occur because of the osmotic pressure arising due to different salt concentrations in intra- and extracellular fluids. Conventional models of cell dehydration during freezing (see e.g. [28] and [29]) describe the change of the cell volume. The cell shape is supposed to be spherical or cylindrical. However, as it is reported by biologists (see e.g. [30]), controlling cell shape is also important for the survival rate of cells. Mathematical models proposed in this section are concerned with the evolution of cell shape depending on the temperature distribution and the amount of intra- and extracellular ice. The study is based on the theory of ice formation in porous media (see [21]) and Stefan-type models (see [14]) describing the motion of the cell membrane due to osmotic flow into and out of the cell. Equations of the Hamilton-Jacobi type are solved numerically using both finite-difference schemes and reachable set methods applied to associated conflict control problems formalized according to [12] and [31]. Examples of modeling cell shape change are presented.

3.1. Ice Formation in Intra- and Extracellular Liquid

The cells of a tissue are surrounded by an extracellular liquid confined in small vessels of an extracellular matrix. Biologists suppose that cells may communicate through these vessels. Since the mechanism and the role of cell-cell interactions are still not well understood (see e.g. [32]), we will consider a simplified model which does not take into account possible cell-cell interactions. For this reason, the extracellular liquid is assumed to be confined in small non-communicating cavities or pores of the extracellular matrix (see a sketch in the left part of Figure 28). Each cell has a membrane that provides a physical separation between the intra- and extracellular environments, which may cause delayed freezing of intracellular liquid. This effect results in a very large stress exerted on the cell membrane. The magnitude of this effect can be approximately estimated as follows:

$$p \approx C^{ice}\,\text{æ} \cdot (1 - \beta_\ell),$$

where p is the pressure, C^{ice} the elasticity of ice, æ the ratio of volume expansion due to the water-to-ice phase transition, β_ℓ the unfrozen water fraction so that $1 - \beta_\ell$ is the ice content. A rough estimate yields $p \approx 1\,\text{Bar}$, which may be dangerous for cell membranes. To reduce this effect, the fluids inside and outside the cell must freeze simultaneously. This can be achieved by lowering the freezing point of the extracellular fluid and optimizing cooling protocols.

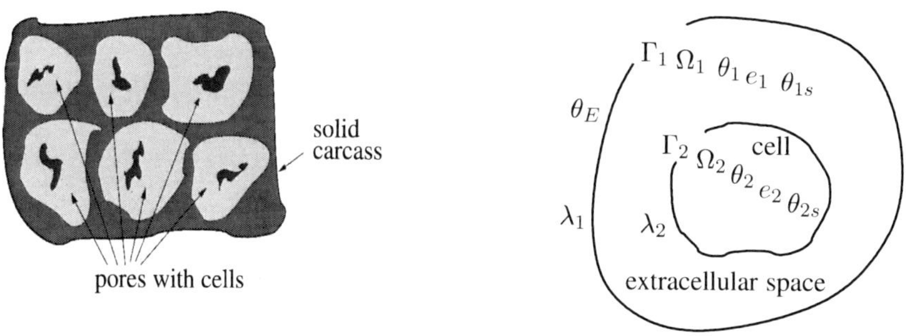

Figure 28. Two-dimensional sketch of an extracellular matrix with pores filled with extracellular fluid and living cells containing intracellular fluid (to the left). Schematic notation of variables and three-dimensional regions related to a single pore (to the right). Note that Γ_1 and Γ_2 are the boundaries of the pore and the cell, respectively, Ω_1 is the region lying between Γ_1 and Γ_2 (extracellular space), and Ω_2 is the region of the cell.

3.1.1. Derivation of Model Equations and Setting of Objectives

In the case were a fluid fills a micro volume, say $\Omega \in R^3$, bounded by a solid wall, the unfrozen water fraction β_ℓ can be computed using the following phase field model (see [21] and Section 2.5):

$$\begin{aligned} &\frac{\partial e(\theta)}{\partial t} - \mathcal{K}\Delta\theta = 0, && -\mathcal{K}\frac{\partial \theta}{\partial \nu}\Big|_{\partial\Omega} = \lambda(\theta - \theta_E)|_{\partial\Omega}, \\ &e(\theta) = \rho C\theta + \rho L \beta_\ell(\theta), && \beta_\ell(\theta) = \phi\left(\frac{L(\theta - \theta_s)}{(T_0 + \theta_s)(T_0 + \theta)}\right), \end{aligned} \tag{37}$$

where θ is the Celsius temperature, $\mathcal{K}$ the heat conductivity coefficient, λ the overall (film) heat transfer coefficient, θ_E the external temperature (outside of Ω), C the specific heat capacity, ρ the density, L the specific latent heat, θ_s the freezing (solidification) point, T_0 the Celsius zero point (273°K). The function $e(\theta)$ denotes the internal energy related to the temperature. The function ϕ is recovered from experimental data.

Consider now a sketch of a pore shown in Figure 28 (to the right). The notations are self-explanatory. It is only to note that λ_1 and λ_2 denote the film heat transfer coefficients of the pore boundary and the cell membrane, respectively, θ_{1s} and θ_{2s} stand for the freezing points of the extracellular and intracellular fluids, respectively. Integrating the energy balance equation of (37) over Ω_1 and Ω_2, using notations

$$\hat{e}_i = \frac{1}{|\Omega_i|}\int_{\Omega_i} e_i dV;\ \hat{\theta}_i = \frac{1}{|\Gamma_i|}\int_{\Gamma_i} \theta_i dS;\ \hat{\theta}_E = \frac{1}{|\Gamma_1|}\int_{\Gamma_1} \theta_E dS;\ \hat{\alpha}_i = \frac{|\Gamma_2|}{|\Omega_i|}\lambda_2;\ \hat{\lambda} = \frac{|\Gamma_1|}{|\Omega_1|}\lambda_1;$$

for mean values, and assuming that $|\Gamma_1|^{-1}\int_{\Gamma_1}\theta_1 dS \approx |\Gamma_2|^{-1}\int_{\Gamma_2}\theta_1 dS$, because θ_1 is almost constant in the small region Ω_1, yield the following coupled system of ordinary differential equations:

$$\frac{d}{dt}\hat{e}_1 = -\hat{\alpha}_1\big[\hat{\theta}_1 - \hat{\theta}_2\big] - \hat{\lambda}\big[\hat{\theta}_1 - \hat{\theta_E}\big], \qquad \frac{d}{dt}\hat{e}_2 = -\hat{\alpha}_2\big[\hat{\theta}_2 - \hat{\theta}_1\big]. \tag{38}$$

The relation between $\hat{e}_i$ and $\hat{\theta}_i$ (see (37)) is given by the formula

$$\hat{e}_i = \rho C\hat{\theta}_i + \rho L\beta_\ell^i(\hat{\theta}_i), \tag{39}$$

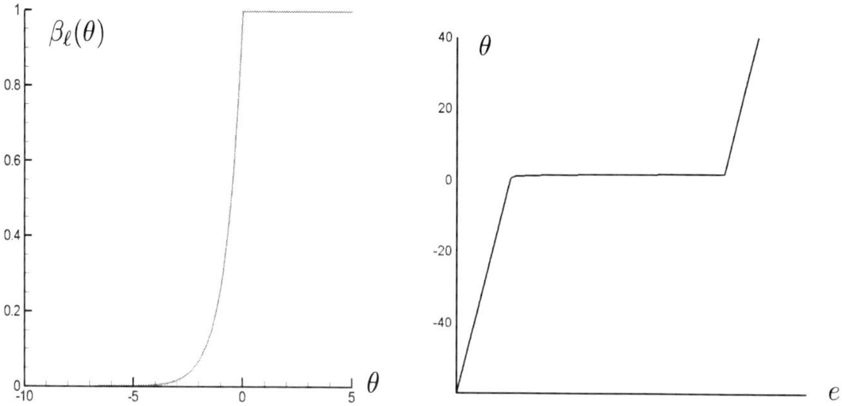

Figure 29. Dependence of the unfrozen water fraction on the temperature (to the left). Graph of the inverse function to $e(\theta)$ (to the right).

where β_ℓ^1 and β_ℓ^2 are defined by the replacement of θ_s with θ_{1s} and θ_{2s}, respectively, in (37).

Note that the derivatives $\partial\hat{e}_i/\partial\hat{\theta}_i$, $i = 1, 2$, are discontinuous (see equation (39) and the left graph of Figure 29). Therefore, the direct form of equations (38) is not appropriate. It is convenient to express the temperatures $\hat{\theta}_1$ and $\hat{\theta}_2$ through the energies $\hat{e}_1$ and $\hat{e}_2$, respectively, to obtain the relation $\hat{\theta}_i = \Theta_i(\hat{e}_i)$, which is the inverse of relation (39) (see the right graph of Figure 29). By doing that, we obtain the following system of ordinary differential equations:

$$\dot{\hat{e}}_1 = -\hat{\alpha}_1\big[\Theta_1(\hat{e}_1) - \Theta_2(\hat{e}_2)\big] - \hat{\lambda}\big[\Theta_1(\hat{e}_1) - \hat{\theta}_E\big], \qquad \dot{\hat{e}}_2 = -\hat{\alpha}_2\big[\Theta_2(\hat{e}_2) - \Theta_1(\hat{e}_1)\big].$$

For simplicity, denote $x = \hat{e}_1$, $y = \hat{e}_2$, $z = \hat{\theta}_E$, $\alpha_i = \hat{\alpha}_i$, $\lambda = \hat{\lambda}$, and consider the following control system:

$$\begin{aligned} \dot{x} &= -\alpha_1[\Theta_1(x) - \Theta_2(y)] - \lambda[\Theta_1(x) - z] - d + v_1 && (=: f_1),\\ \dot{y} &= -\alpha_2[\Theta_2(y) - \Theta_1(x)] + v_2 && (=: f_2),\\ \dot{z} &= u && (=: f_3). \end{aligned} \tag{40}$$

Here, z is the temperature outside of the pore (chamber temperature), u is the cooling rate, v_1, v_2 are disturbances interpreted as data errors. The control variable u (cooling rate) is restricted by $|u| \leq \mu$, the disturbances v_1, v_2 are bounded by $|v_1| \leq \nu$, $|v_2| \leq \nu$. Since the zero initial value for z is always assumed, the constant d is introduced to emulate nonzero initial values of z (nonzero initial chamber temperatures).

According to the meaning of the functions β_ℓ^i, $i = 1, 2$, exact simultaneous freezing of the extracellular and intracellular liquids can be expressed as the vanishing of the following functional:

$$J = \int_0^{t_f} \gamma^2(x(t), y(t))dt, \tag{41}$$

where

$$\gamma(x, y) = |\beta_\ell^1(\Theta_1(x)) - \beta_\ell^2(\Theta_2(y))|$$

estimates the difference of the ice fractions in the extra- and intracellular regions.

Consider differential game (40) and (41) assuming that the objective of the control u is to minimize the functional J, whereas the objective of the disturbance is opposite. Moreover, the trajectories should remain in a state constraint set N given by the relation

$$N := \{(t, x, y, z) : t \in [0, t_f],\ \ \omega(t, x, y, z) \leq 0\}, \tag{42}$$

where $\omega : [0, t_f] \times R^3 \to R$ is a given function.

It is known, that the value function of a differential game completely defines an optimal feedback control. The value function can be computed as a viscosity solution [13] to an appropriate HJBI equation, and the optimal feedback control can be designed by applying the procedure of extremal aiming (see [12]).

3.1.2. Dynamic Programming Method for Optimization of Cooling and Freezing Protocols

The game (40)–(42) is formalized as in [12] and [33], which ensures the existence of the value function V. The next important result proved in [34] is that V coincides with a unique viscosity solution of the HJBI equation

$$V_t + \mathsf{H}(x, y, z, V_x, V_y, V_z) = 0, \tag{43}$$

where the Hamiltonian H and the terminal condition are defined as

$$\mathsf{H}(x, y, z, p_1, p_2, p_3) = \max_{|v_1|,|v_2| \leq \nu} \min_{|u| \leq \mu} \sum_{i=1}^{3} p_i f_i + \gamma^2(x, y), \quad V(t_f, x, y, z) = 0.$$

In the next section, an approximation scheme for solving this HJBI equation is discussed and a convergence result is given.

3.1.3. Numerical Methods for Solving Optimal Control Problems

Let h_x, h_y, h_z, and τ be space and time discretization step sizes. Assume that $N = t_f/\tau$ is integer and introduce the notation

$$V^n(x_i, y_j, z_k) = V(n\tau, ih_x, jh_y, kh_z), \quad \omega^n(x_i, y_j, z_k) = \omega(n\tau, ih_x, jh_y, kh_z),$$

where $n \in \overline{0, N}$. Therefore, V^n and ω^n are grid functions defined on the above introduced spatial grid. Let $V^n \rightarrow \Pi(V^n; \tau, h_x, h_y, h_z)$ be an operator that maps grid functions into grid functions and can be naturally extended to continuum functions. An example of such an operator is given by the formula

$$\Pi(V^n; \tau, h_x, h_y, h_z) = V^n(x_i, y_j, z_k) + \tau \mathsf{H}(x_i, y_j, z_k, V_x^n, V_y^n, V_z^n),$$

where V_x^n, V_y^n, V_z^n denote finite difference approximations (left, right, central and etc.) of the corresponding partial derivatives.

Definition 3.1. *The operator Π is monotone if, for any grid functions V and W, the following implication holds point-wise:*

$$V \leq W \Rightarrow \Pi(V; \tau, h_x, h_y, h_z) \leq \Pi(W; \tau, h_x, h_y, h_z).$$

Definition 3.2. *The operator Π has the generator property if there exist a, b, $c > 0$ such that the estimate*

$$\left| \frac{\Pi(\phi; \tau, a\tau, b\tau, c\tau)(\vec{r}) - \phi(\vec{r})}{\tau} - \mathsf{H}(\vec{r}, D\phi(\vec{r})) \right| \leq C\big(1 + \|D\phi\| + \|D^2\phi\|\big)\tau \tag{44}$$

holds for all $\phi \in C_b^2(\mathcal{D})$, $\vec{r} = (x, y, z) \in \mathcal{D}$. Here $\mathcal{D} \subset R^3$ is a large domain containing all states of the game, $C_b^2(\mathcal{D})$ is the space of twice continuously differentiable functions defined on $\mathcal{D}$ and bounded together with their two derivatives, $\| \cdot \|$ denotes the point-wise maximum norm, $D\phi$ and $D^2\phi$ denote the gradient and the Hessian matrix of ϕ, respectively.

Theorem 3.1. *Assume that the operator $\Pi(\cdot; \tau, a\tau, b\tau, c\tau)$ is monotone for any $\tau > 0$ and satisfies the estimate (44), then the time-space grid function $\{V^0, V^1, ..., V^N\}$ obtained using the relation*

$$V^{n-1} = \max\{\Pi(V^n; \tau, a\tau, b\tau, c\tau),\ \omega^n\}, \quad n = 1, ..., N, \quad V^N = 0 \tag{45}$$

converges point-wise to a unique viscosity solution of (43) and, therefore, to the value function of the differential game (40)–(42) *as* $\tau \to 0$*. The convergence rate is* $\sqrt{\tau}$.

Proof. First, note that the problem with the cost functional (41) and state constraint (42) can be reduced to the case considered in [35] by introducing the auxiliary equation $\dot{\eta} = \gamma^2(x, y)$ and considering the cost functional of the form

$$J = \max\left\{\eta(t_f), \max_{t\in[t_0,t_f]} \omega(t, x(t), y(t), z(t))\right\}.$$

Therefore, we have a particular case of a more general problem considered in [35]. □

Remark 3.1. *Theorem 3.1 refers only to the monotonicity and generator properties of the operator* Π*. Really, some secondary properties must hold to provide the convergence (see [35] and [36]). It is easy to prove that these properties obviously hold for an upwind operator* Π *presented in the next section.*

3.1.4. Upwind Time Step Operator

We will consider an upwind time step operator proposed in [37] and prove that it is monotone and possesses the generator property, which proves the convergence result.

Assume that the right-hand-sides f_i, $i = 1, 2, 3$, are now arbitrary functions of the state (x, y, z). Let $u \in P \subset R^p$ and $v \in Q \subset R^q$. Denote $a^+ = \max(a, 0)$ and $a^- = \min(a, 0)$. The operator introduced in [37] assumes the following approximations of the spatial derivatives:

$$V_x^n \cdot f_1 = p_1^R \cdot f_1^+ + p_1^L \cdot f_1^-, \quad V_y^n \cdot f_2 = p_2^R \cdot f_2^+ + p_2^L \cdot f_2^-, \quad V_z^n \cdot f_3 = p_3^R \cdot f_3^+ + p_3^L \cdot f_3^-,$$

where f_1, f_2, f_3 are computed at (x_i, y_j, z_k); p_1^R, p_2^R, p_3^R and p_1^L, p_2^L, p_3^L the right and left divided differences, respectively, defined as

$$p_1^R = [V^n(x_{i+1}, y_j, z_k) - V^n(x_i, y_j, z_k]/h_x,$$
$$p_2^R = [V^n(x_i, y_{j+1}, z_k) - V^n(x_i, y_j, z_k]/h_y,$$
$$p_3^R = [V^n(x_i, y_j, z_{k+1}) - V^n(x_i, y_j, z_k]/h_z,$$

$$p_1^L = [V^n(x_i, y_j, z_k) - V^n(x_{i-1}, y_j, z_k]/h_x,$$
$$p_2^L = [V^n(x_i, y_j, z_k) - V^n(x_i, y_{j-1}, z_k]/h_y,$$
$$p_3^L = [V^n(x_i, y_j, z_k) - V^n(x_i, y_j, z_{k-1}]/h_z.$$

Finally, the operator is given by

$$\Pi(V^n;\tau,h_x,h_y,h_z)(x_i,y_j,z_k) = V^n(x_i,y_j,z_k) + \tau\gamma^2(x_i,y_j,z_k)$$
$$+\tau \max_{v\in Q}\min_{u\in P}\left(p_1^R\cdot f_1^+ + p_1^L\cdot f_1^- + p_2^R\cdot f_2^+ + p_2^L\cdot f_2^- + p_3^R\cdot f_3^+ + p_3^L\cdot f_3^-\right). \tag{46}$$

Lemma 3.1. *Let M be the bound of the right hand side of the control system. If $a,b,c\geq M\sqrt{3}$, then the operator $\Pi(\cdot;\tau,a\tau,b\tau,c\tau)$ given by* (46) *is monotone. The generator property* (44) *holds for any fixed a,b,c.*

The proof is the same as in Lemmas 3.2 and 3.3 of Section 3.2.3 (see also [38] and [39]). Thus, the operator (46) satisfies the conditions of Theorem 3.1.

Remark 3.2. *If the functions $f_i, i=1,2,3$, are linear in u and v at each fixed state (x,y,z), then the operation $\max_{v\in Q}\min_{u\in P}$ appearing in the definitions of the operator (46) can be replaced by $\max_{v\in \mathit{ext}\,Q}\min_{u\in \mathit{ext}\,P}$, where "ext" returns the set of the extremal points. In particular, "$\max\min$" can be computed over the set of vertices, if P and Q are polyhedrons. The monotonicity holds independently of the structure of the sets P and Q. To prove the generator property, it is sufficient to observe that the Hamiltonian is equal to that computed using the sets ext P and ext Q whenever the assumption of linearity in u and v at each fixed state point holds.*

Note that this remark is very important for numerical implementations of the operator (46) because the operation "$\max\min$" is applied to a function which is nonlinear and neither convex nor concave in u and v.

3.1.5. Control Procedure

In this section, the computation of optimal controls for system (40) using the procedure of extremal aiming (see [12]) is described.

Let ε be a small positive number, t_n the current time instant. Consider the neighborhood

$$\mathcal{U}_\varepsilon = \{(x,y,z)\in R^3 :\ |x-x(t_n)|\leq\varepsilon,\ |y-y(t_n)|\leq\varepsilon,\ |z-z(t_n)|\leq\varepsilon\}$$

of the current state $\big(x(t_n), y(t_n), z(t_n)\big)$ of system (40). By searching through all grid points $(x_i, y_j, z_k) \in \mathcal{U}_\varepsilon$, find a point $(x_{i_*}, y_{j_*}, z_{k_*})$ such that

$$V^n(x_{i_*}, y_{j_*}, z_{k_*}) = \min_{(x_i, y_j, z_k) \in \mathcal{U}_\varepsilon} V^n(x_i, y_j, z_k).$$

The current control $u(t_n)$ which is supposed to be applied on the next time interval $[t_n, t_n + \tau]$ is computed from the condition of maximal projection of the system velocity (f_1, f_2, f_3) onto the direction of the vector $\big(x_{i_*} - x(t_n),\ y_{i*} - y(t_n),\ z_{i_*} - z(t_n)\big)$, i.e.

$$u(t_n) = \arg\max_{|u| \le \mu} \Big(\big(x_{i_*} - x(t_n)\big) f_1 + \big(y_{i_*} - y(t_n)\big) f_2 + \big(z_{i_*} - z(t_n)\big) f_3\Big).$$

It is clear that the value of the control will be either μ or $-\mu$.

3.1.6. Simulation of Cell Freezing with Accounting for Balance of Ice Formation, Supercooling Effects, and Growth of Dendrite Seeds

Balance of Ice Formation with Accounting for Supercooling Effects

Consider first the following two-dimensional variant of the control system:

$$\begin{aligned} \dot{x} &= -\alpha_1[\Theta_1(x) - \Theta_2(y)] - \lambda[\Theta_1(x) - u] - d + v_1 \\ \dot{y} &= -\alpha_2[\Theta_2(y) - \Theta_1(x)] + v_2, \end{aligned} \tag{47}$$

which corresponds to the assumption $z \equiv u$ (infinite cooling rate). The values of the coefficients and bounds on the control and disturbances used for all simulations are: $\alpha_1 = \alpha_2 = 0.1$, $\lambda = 2$, $d = 2$, $\mu = 4$, $\nu = 0.2$. The notation $\beta_i^i := 1 - \beta_\ell^i$, $i = 1, 2$, is used for the ice fractions.

We start with the case where the intra- and extracellular fluids have the same freezing points, i.e. $\theta_{1s} = \theta_{2s}$. The plots of the temperatures in the cell and pore regions versus time are given in Figure 30a. Figure 30b shows that the control fails to balance the ice fractions in the pore and in the cell.

The next simulation (see Figure 31) shows the case of different freezing points for the pore and the cell: $\theta_{1s} - \theta_{2s} = -13$°C. Thus, the freezing point of the extracellular fluid is lowered, e.g. by adding a cryoprotector. Now, we can freeze the intracellular fluid using temperatures laying above the freezing point of the extracellular liquid, which makes possible simultaneous freezing.

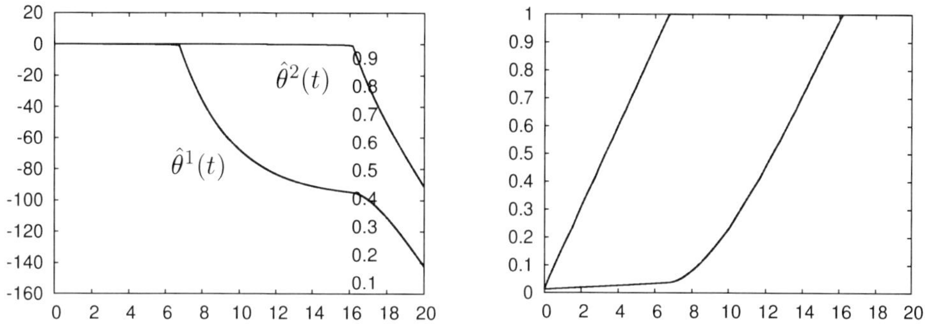

(a) Temperatures in the extracellular space and in the cell versus time; latent heat plateaus are present

(b) Ice fraction in the extracellular space and in the cell versus time

Figure 30. The case of infinite cooling rate, and $\theta_{1s} = \theta_{2s}$.

Figure 32 stands for the same setting as Figure 31 but in the case of finite cooling rate (see equations (40)).

The effect of supercooling of the intracellular fluid manifested as a kink in the dependence of the temperature on the internal energy (see Figure 33a for an experimentally measured curve) can be better accounted for using an idealized curve presented in Figure 33b.

Figures 34a, 34b, and 34c give the graph of the value function, the temperatures, and the ice contents, respectively. The control with very frequent switches which is able to equalize the ice contents inside and outside the cell (see Figure 34c) is shown in Figure 34d.

Computation results related to system (40) that models finite cooling rates are presented in Figs. 35a–35d. Also in this more realistic case, the control ensures keeping the ice contents in the cell and in the extracellular space very close to each other. Note that the computation here is performed in three dimensions. The section of the graph of the value function by the plane $z = 0$ is shown in Figure 35a. For this simulation, we have also computed the integral $\int_0^{t_f} |c_{\text{in}}(t) - c_{\text{out}}(t)|\beta_\ell^1\beta_\ell^2 dt$ that estimates the outflow from the cell during freezing. The value obtained using the optimal cooling protocol is approximately 2.7 times less than the one produced by a heuristic cooling regime. We do not present the graph of the realization of the heuristic regime because it is very similar to that shown in Figure 35d with the exception that there is only one switching (from -8 to 0) that occurs at $t = 1.512$. Note that the final switching point (from 0 to -8) of the optimal cooling protocol occurs at $t = 60.8$, i.e. shortly before complete freezing of the intra- and extracellular fluids.

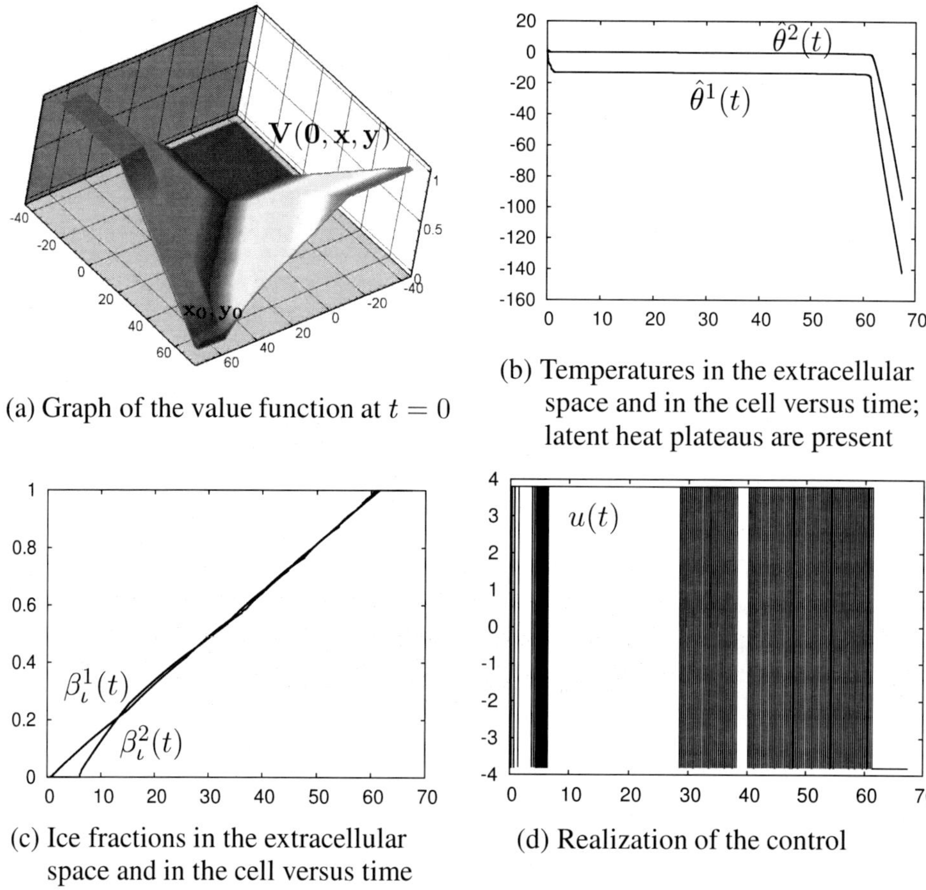

(a) Graph of the value function at $t = 0$

(b) Temperatures in the extracellular space and in the cell versus time; latent heat plateaus are present

(c) Ice fractions in the extracellular space and in the cell versus time

(d) Realization of the control

Figure 31. The case of infinite cooling rate, $\theta_{1s} - \theta_{2s} = -13°\text{C}$.

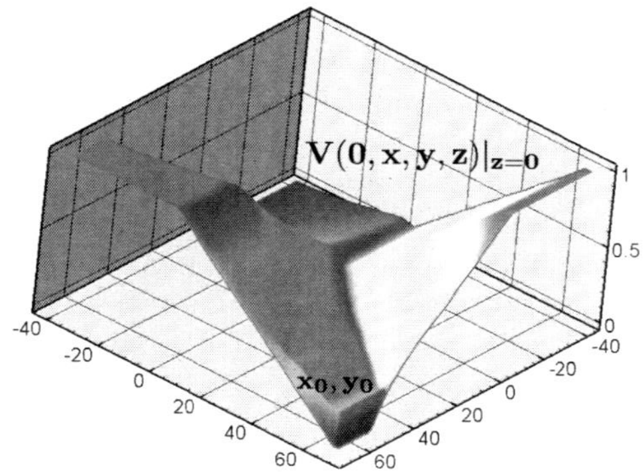

(a) Graph of the value function at $t = 0,\ z = 0$

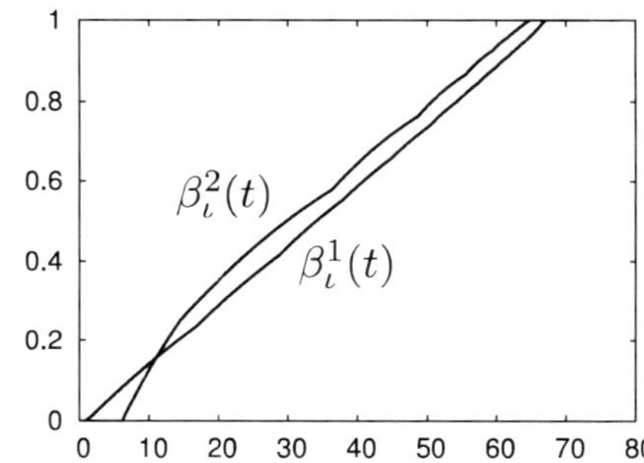

(b) Ice fractions in the extracellular space and in the cell versus time

Figure 32. The realistic case of finite cooling rate (three dimensions), $\theta_{1s} - \theta_{2s} = -13°C$.

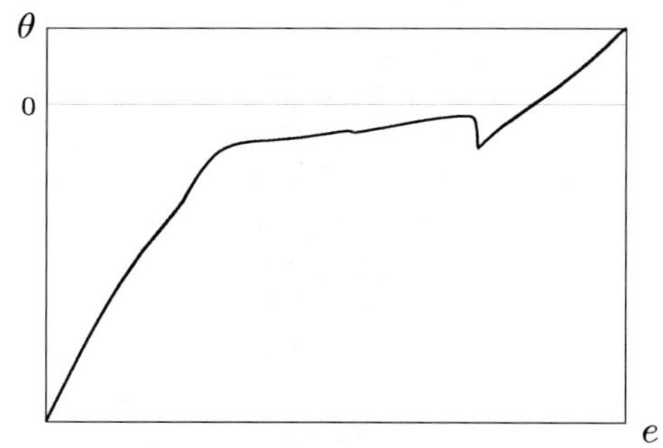

(a) Experimentally measured dependence

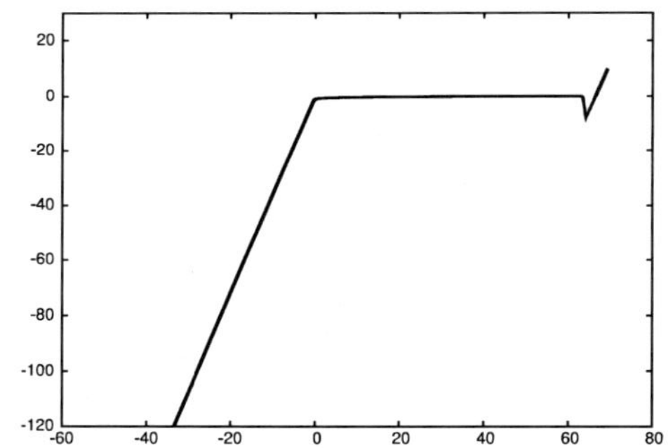

(b) Idealized curve

Figure 33. Supercooling effect in the intracellular fluid displays itself as a kink in the dependence of the temperature on the internal energy.

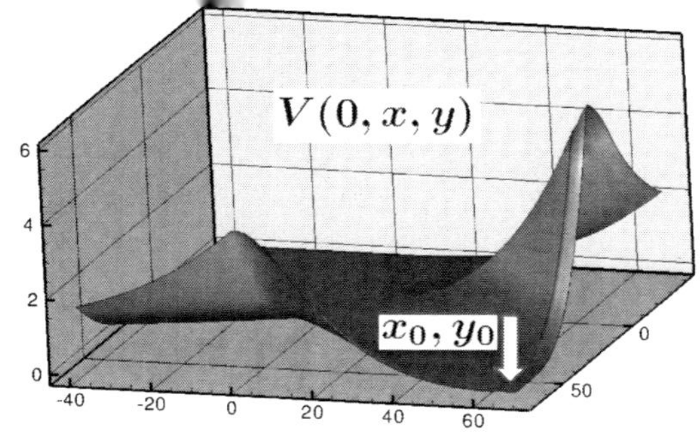

(a) Graph of the value function

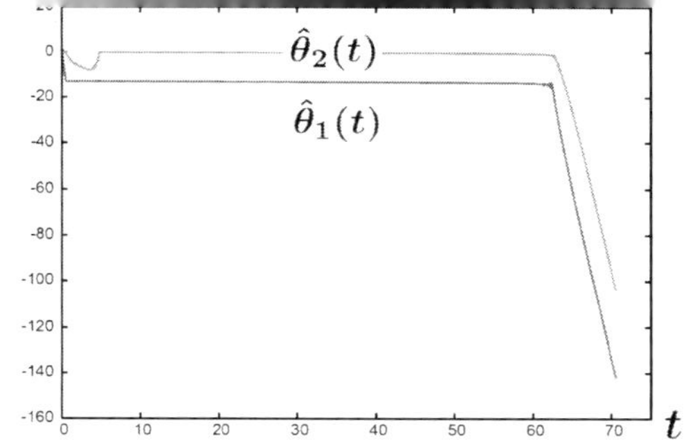

(b) Temperatures in the cell and in the pore versus time

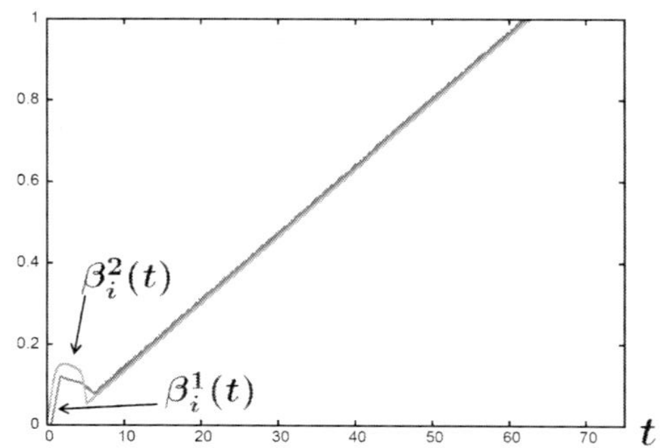

(c) Ice content in the cell and in the pore versus time

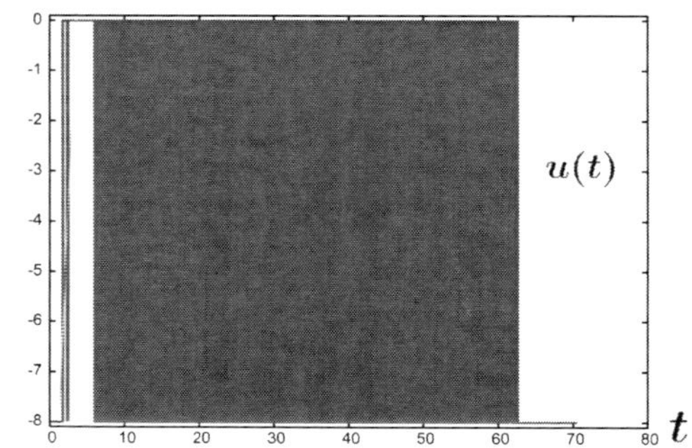

(d) Realization of the control

Figure 34. The case of infinite cooling rate; $\theta_{1s} - \theta_{2s} = -13°\text{C}$.

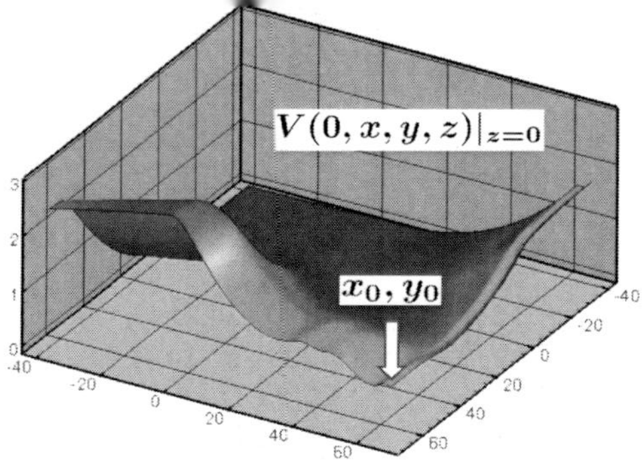

(a) Section of the graph of the value function by the plane $z = 0$

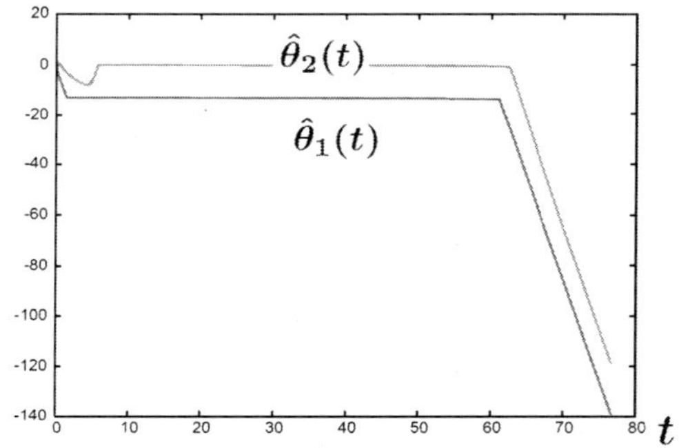

(b) Temperatures in the cell and in the pore versus time

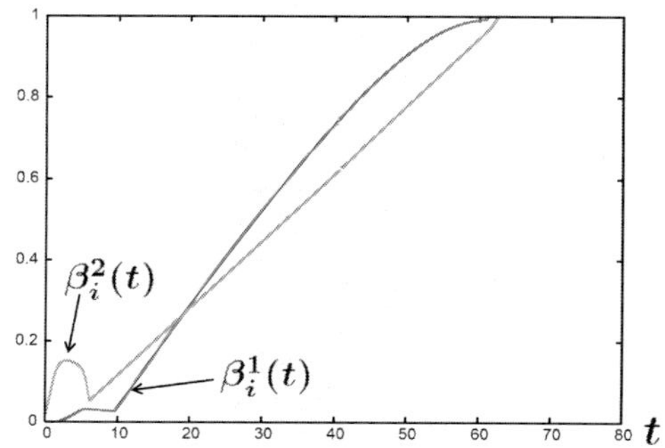

(c) Ice content in the cell and in the pore versus time

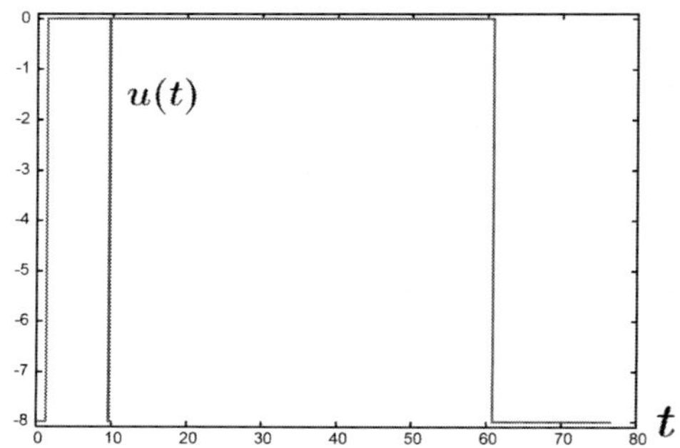

(d) Realization of the control

Figure 35. The case of finite cooling rate; $\theta_{1s} - \theta_{2s} = -13°C$.

Balance of Ice Formation with Accounting for Dendrite Growth

To account for the growth of dendrite seeds, consider internal energies (see Section 3.1.1) outside the cell,

$$\hat{e}_1(\hat{\theta}_1) = \rho C\hat{\theta}_1 + \rho L\beta_\ell^1(\hat{\theta}_1),$$

and inside the cell,

$$\hat{e}_2(\hat{\theta}_2) = \rho C\hat{\theta}_2 + \rho L\beta_\ell^2(\hat{\theta}_2) + \rho D\kappa\beta_\ell^2(\hat{\theta}_2),$$

where the last term is treated as a dendrite generation energy (D is the specific latent heat of dendrite growth). The value κ is computed as follows:

$$\kappa = \kappa_0(\dot{\hat{e}}_2)^- \;=\; -\kappa_0\alpha_2(\hat{\theta}_2 - \hat{\theta}_1)^+,$$

where $(a)^- = \min\{a, 0\}$ and $(a)^+ = \max\{a, 0\}$. We solve equations for $\hat{e}_1$ and $\hat{e}_2$ numerically to find $\Theta_1(x)$ and $\Theta_2(x, y)$, remember the notation $x = \hat{e}_1$ and $y = \hat{e}_2$, and put them into the system (40).

The objective of the control is to minimize the generation of dendrite seeds along with the balance of ice formation so that the following extended functional is considered

$$\begin{aligned} J = &\int_0^{t_f} \left|\beta_\ell^1(\Theta_1(x(\tau))) - \beta_\ell^2(\Theta_2(y(\tau)))\right|^2 d\tau + \\ &\kappa_0\rho D\alpha_2 \int_0^{t_f} \big(\Theta_2(y(\tau)) - \Theta_1(x(\tau))\big)^+ \beta_\ell^2(\Theta_2(y(\tau)))d\tau. \end{aligned} \tag{48}$$

Figure 36 shows results of numerical simulation related to system (40) and functional (48). Additionally, the state constraint $z \leq 2°\text{C}$ (remember that $z = \hat{\theta}_E$, the scaled temperature outside the pore) is imposed. It was observed that 30% less dendritic seeds are formed if functional (48) is used instead of functional (41).

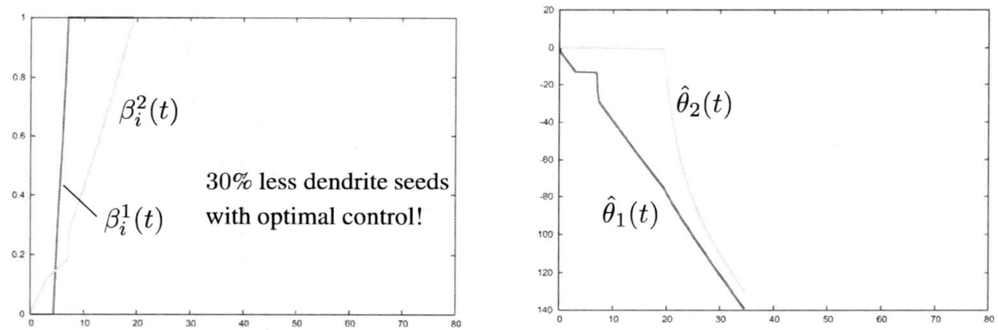

(a) (b)

Figure 36. Minimization of the growth of dendritic seeds at the freezing stage: (a) Ice fraction versus time; (b) Realizations of the temperatures inside and outside the cell.

Remember that the central point of the control design is the computation of the value function using the upwind grid method given by (45) and (46). Numerical experiments show a very nice property of this method: the noise usually coming from the boundary of the grid region is absent. The examples are calculated on a Linux computer admitting 64 GB memory and 32 threads. The coefficient of the parallelization is equal to 0.7 per thread (23 times speedup totally). The grid size in three dimensions is 300^3, the number of time steps is 30000 (see the restrictive relation between the space and time step sizes given by Lemma 3.1). The runtime is approximately 60 min. In the case of two dimensions, the runtime is several minutes. Figures 31a, 32a, and 35a demonstrate graphs of the value function.

3.1.7. Simulation of Cell Thawing with Optimization of the Osmotic Inflow and Accounting for the Dendrite Growth

Optimization of the Osmotic Inflow

First note that, comparing to the freezing procedure, the dependence of the unfrozen water fraction on the temperature is modified as it is shown in Figure 37a. The functional to minimize expresses the amount of liquid moving into the cell, which is proportional to the difference between the concentration of the physiological salt solution and salt concentration in the cell

$$J = \alpha \int\limits_0^{t_f} \left| c_0 - g\Big(\frac{m_0}{W_0\, \beta_\ell^2(\Theta_2(y(\tau)))} \Big) \right| d\tau. \tag{49}$$

Here, c_0 is the concentration of the physiologic salt solution, m_0 the salt amount in the cell, W_0 the initial cell volume, g the function defining the salt concentration in the cell. Thus, $W_0\beta_\ell^2$ is the unfrozen water volume inside the cell, and $g(m_0/(W_0\beta_\ell^2))$ the salt concentration in the cell (see Figure 37b). Additionally, the state constraints

$$-50°\mathrm{C} \leq z \leq 40°\mathrm{C} \text{ and } \Theta_2(y) \leq 20°\mathrm{C} \tag{50}$$

are imposed.

Figure 38 demonstrate simulation results related to system (40) and functional (49). In the simulation process, the amount of fluid transferred into the cell was computed. This value is 65% less in the case of the optimal control than in the case of applying a reasonable (not optimal) control.

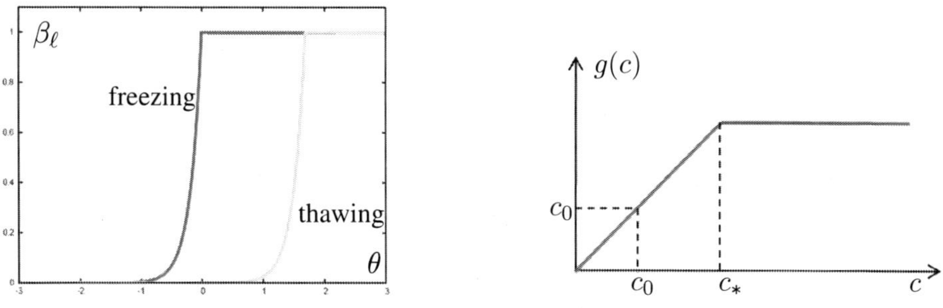

(a) (b)

Figure 37. Cell thawing: (a) Unfrozen liquid fraction versus temperature; (b) Function defining variation of the salt concentration in the cell.

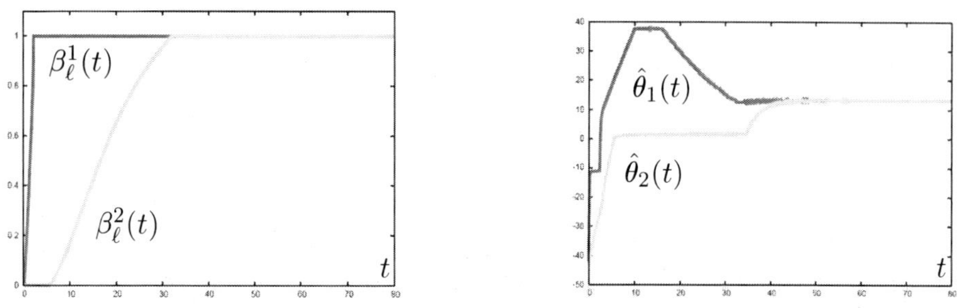

(a) (b)

Figure 38. Optimization of osmotic flow into the cell at the thawing stage: (a) Unfrozen fluid fraction versus time; (b) Realization of the temperatures inside and outside the cell.

Accounting for The Dendrite Growth

Dendrite seeds formed during freezing, can form large ice crystals at the thawing stage, which may be dangerous for cells. To take into account the growth of dendrites, the following functional is considered:

$$J = \alpha \int_0^{t_f} \left| c_0 - g\Big(\frac{m_0}{W_0\, \beta_\ell^2(\theta_2(y(\tau)))}\Big) \right| d\tau + \max_{t\in[0,t_f]} \big[(1 - \beta_\ell^2(\theta_2))p(t)\beta_\ell^2(\theta_2)\big]. \tag{51}$$

The first term in (51) coincides with (49), and the second one expresses the amount of ice formed due to aggregation of dendrite seeds according to the Poisson law (see [40])

$$p(t) = 1 - e^{-\lambda(\theta,\Delta E, D(\theta))t}.$$

The nucleation rate λ is a function of the temperature, the activation energy, and the diffusive mobility of dendritic seeds, respectively. In the simulation, the nucleation rate λ is supposed to be constant.

Since both terms in the functional are non-antagonistic, it is clear that rapid thawing is preferable. To prevent excessive warming, state constraints (50) are applied.

Two simulations of system (40) with state constraints (50) were performed: the first one with functional (49), and the second one with functional (51). The comparison of the simulation results shows that 7% less dendrites are formed in the second case.

3.2. Mathematical Models of Dehydration and Rehydration of Cells

This section is devoted to the modeling of the change of cell shape and volume caused by the dehydration and rehydration processes. Each biological cell is located inside a pore or channel filled with a saline solution called extracellular fluid. The cell interior is separated from the outer liquid by a cell membrane whose structure ensures a very good permeability for water. This makes possible easy water inflow and outflow due to the osmotic pressure arising due to different salt concentrations in intra- and extracellular fluids.

3.2.1. Dehydration of Cells

In the freezing phase, the mechanism of the osmotic effect is the following. Ice formation occurs initially in the extracellular solution. Since ice is practically free of salt, the water-to-ice phase change results in the increase of the salt concentration (c_{out}) in the remaining extracellular liquid. The osmotic pressure forces the outflow of water from the cell to balance the intracellular (c_{in}) and extracellular (c_{out}) salt concentrations. Modeling of cell shrinkage is based on free boundary problem techniques. The main relation here is the so-called Stefan condition: $\mathcal{V} = \alpha(c_{\text{out}} - c_{\text{in}})$, where $\mathcal{V}$ is the normal velocity of the cell boundary (directed to the cell interior), and the right-hand-side represents the osmotic flux that is proportional to the difference of the salt concentrations. The coefficient α is the product of the Boltzmann constant, the temperature, and the hydraulic conductivity of the membrane (see e.g. [29]). Note that α is practically constant in our case. The extracellular salt concentration c_{out} depends on the unfrozen fraction β_ℓ of the extracellular liquid (see the phase field model (37)). Remember that the function $\beta_\ell(\theta)$ is a constitutive material law that, e.g. in frozen-soil science, is measured directly by nuclear magnetic resonance.

The intracellular and extracellular salt concentrations are estimated using the mass conservation law as follows:

$$c_{\text{in}} = c_{\text{in}}^0 W_c^0 / W_c, \quad c_{\text{out}} = c_{\text{out}}^0 W^0 / W,$$

where W_c^0 and W_c are the initial and current cell volumes, respectively, W^0 and W are the initial and current volumes of the unfrozen part of the pore. The current volume W at time t is computed as

$$W(t) = \int\limits_{W^0} \beta_\ell(\theta(t,x))dx,$$

where the distribution of the temperature $\theta(t,x)$ is found from the phase field model (37). A typical form of the function $W(t)$ is shown in Figure 39.

The cell region $\Sigma(t)$ is searched as the level set of a function $\Psi(t,x)$, i.e.

$$\Sigma(t) = \{x\colon\ \Psi(t,x) \le 1\}, \qquad x \in R^3 \text{ (or } R^2\text{)}.$$

Assuming that the cell boundary propagates with the normal velocity $\mathcal{V}$ yields the following Hamilton-Jacobi equation for the function $\Psi(t,x)$:

$$\Psi_t - \alpha(c_{\text{out}} - c_{\text{in}})|\nabla\Psi| = 0, \quad \Psi(0,x) = \inf\{\lambda > 0\colon\ x \in \lambda \cdot \Sigma(0)\}. \quad (52)$$

Here $|\nabla\Psi|$ denotes the Euclidean norm of the gradient.

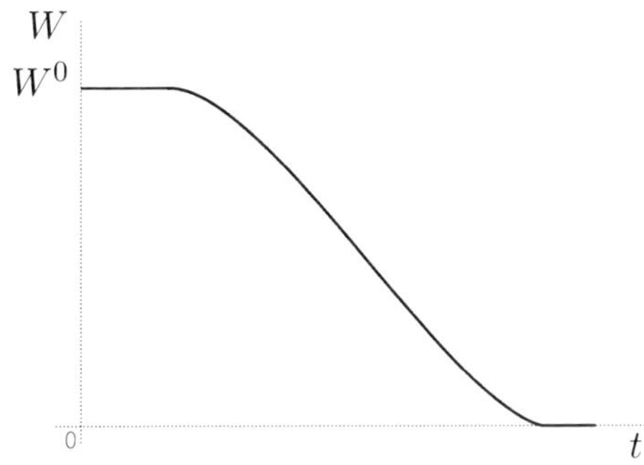

Figure 39. Evolution of the volume of the unfrozen extracellular liquid versus time.

3.2.2. Rehydration of Cells

In the thawing phase, the osmotic effect results in the inflow of water into cells and therefore causes cell swelling. We use the following mass conservation law for the salt content:

$$W_c^0\, c_{\text{in}}^0 = W_s\, c_{\text{in}}^0 + W_\ell\, c_{\text{in}}, \quad W_c = W_s + W_\ell, \tag{53}$$

where W_c^0 is the initial volume of the frozen cell, W_c the current volume of the cell, W_s and W_ℓ are volumes of the frozen and unfrozen parts of the cell, respectively, c_{in}^0 and c_{in} the salt concentrations in the frozen and unfrozen parts of the cell, respectively. Relation (53), yields:

$$c_{\text{in}} = c_{\text{in}}^0 \left(1 + (W_c^0 - W_c)/W_\ell\right).$$

We assume that W_c is calculated from the current cell shape, and

$$W_\ell(t) \approx \int\limits_{W_c^0} \beta_\ell\left(\theta(t,x)\right)dx,$$

where the function β_ℓ shows now the volume fraction of unfrozen intracellular fluid. The salt concentration c_{out} outside the cell is supposed to be a constant.

Admitting that the propagation velocity $\mathcal{V}$ of the cell boundary is proportional to the difference of the concentrations c_{in} and c_{out}, one arrives at an equation of the form of (52).

3.2.3. Finite-Difference Scheme for Solving Hamilton-Jacobi Equation

Hamilton-Jacobi equation (52) is solved numerically using a finite difference scheme for finding viscosity solutions. The aim of this section is to describe a proper finite-difference scheme. Let the symbol "$|\ |$" denotes the absolute value in the case of scalar or the Euclidian norm in the case of vector.

Consider a general form of the Hamilton-Jacobi equation (initial value problem) and assume for simplicity that the spatial variable is three dimensional:

$$\Psi_t + \mathsf{H}(t, x, \Psi_x) = 0, \qquad \Psi(0, x) = \sigma(x). \tag{54}$$

Here $x = (x_1, x_2, x_3) \in R^3$, $t \in [0, \infty)$, $\Psi : [0, \infty) \times R^3 \to R$, and $\sigma : R^3 \to R$ is some given function. Assume that the Hamiltonian H is defined as

$$\mathsf{H}(t, x, p) = \max_{v \in Q} \min_{u \in P} \langle p, f(t, x, u, v) \rangle,$$

where f is the right-hand side of a general conflict control system:

$$\dot{x} = f(t, x, u, v), \quad x \in R^3, \quad u \in P \subset R^p, \quad v \in Q \subset R^q, \quad p, q \leq 3. \tag{55}$$

The function f is assumed to be uniformly continuous on $[0, \infty) \times R^3 \times P \times Q$, bounded and Lipschitz continuous in t, x. The function σ is bounded and Lipschitz continuous in x.

Let τ, h_i, $i = 1, 2, 3$, be time and space discretization steps. Similar to Section 3.1.3, introduce the notation

$$\Psi^n(x_1^i, x_2^j, x_3^k) = \Psi(t_n, ih_1, jh_2, kh_3), \quad t_n = n\tau$$

and consider a finite-difference scheme

$$\Psi^{n+1}(x_1^i, x_2^j, x_3^k) = \Psi^n(x_1^i, x_2^j, x_3^k) - \tau \mathsf{H}(t_n, x_1^i, x_2^j, x_3^k, \Psi^n_{x_1}, \Psi^n_{x_2}, \Psi^n_{x_3})$$

$$\Psi^0(x_1^i, x_2^j, x_3^k) = \sigma(x_1^i, x_2^j, x_3^k).$$

Here, $\Psi^n_{x_1}, \Psi^n_{x_2}, \Psi^n_{x_3}$ denote finite difference approximations (left, right, central and etc.) of the corresponding partial derivatives. Note that the $n+1$ th function is computed on the base of the n th function, and the Hamiltonian appears with the sign "-" in contrast to Section 3.1.3, because the initial value problem is studied here.

The scheme can be considered as the successive application of an operator Π to the grid functions Ψ^n:

$$\Psi^{n+1} = \Pi(\Psi^n; t_n, \tau, h_1, h_2, h_3).$$

Note that such an operator can be naturally extended to continuum functions. Let us remember the monotonicity and generator properties (compare with Section 3.1.3) of the operator Π. The definition of monotonicity is just the same as in Section 3.1.3. The generator property is formulated similar to that from Section 3.1.3 but with the sign "+" in front of the Hamiltonian, i.e.

$$\left| \frac{\Pi(\phi; t, \tau, a_1\tau, a_2\tau, a_3\tau)(x) - \phi(x)}{\tau} + \mathsf{H}(t, x, D\phi(x)) \right| \leq C\big(1 + \|D\phi\| + \|D^2\phi\|\big)\tau$$

for smooth functions ϕ (see Section 3.1.3).

Theorem 3.2. *Assume that the operator $\Pi(\cdot; t, \tau, a_1\tau, a_2\tau, a_3\tau)$ is monotone for any $t,\ \tau\ >\ 0$, and satisfies the generator property, then the grid function obtained by the procedure*

$$\Psi^{n+1} = \Pi(\Psi^n; t, \tau, a_1\tau, a_2\tau, a_3\tau), \quad n = 0, 1, ..., \qquad \Psi^0 = \sigma,$$

converges point-wise to a viscosity solution of Hamilton-Jacobi equation (54) *as $\tau \to 0$, and the convergence rate is $\sqrt{\tau}$.*

The proof of this theorem can be found in [36].

Consider an upwind operator similar to that given in Section 3.1.4 (distinctions arise because of the initial value problem formulation). Namely,

$$\Pi(\Psi^n; t_n, \tau, h_1, h_2, h_3)(x_1^i, x_2^j, x_3^k) = \Psi^n(x_1^i, x_2^j, x_3^k) - \tau \max_{v\in Q} \min_{u\in P} \sum_{m=1}^{3} (p_m^L \cdot f_m^+ + p_m^R \cdot f_m^-), \tag{56}$$

where f_1, f_2, f_3 are the right hand sides of system (55) computed at $(t_n, x_1^i, x_2^j, x_3^k, u, v)$; p_1^R, p_2^R, p_3^R and p_1^L, p_2^L, p_3^L the right and the left divided differences, respectively. The arguments $t_n, x_1^i, x_2^j, x_3^k, u, v$ of f_m^- and f_m^+ are omitted for brevity.

Prove that the operator Π satisfies the conditions of Theorem 3.2. The following lemmas hold.

Lemma 3.2. (Monotonicity). *Let M be the bound of $|f|$. If $a_1, a_2, a_3 \geq M\sqrt{3}$, then the operator $\Pi(\cdot; t, \tau, a\tau, b\tau, c\tau)$ given by* (56) *is monotone.*

Proof. Suppose $\Psi \le \Phi$ and show that

$$\Pi(\Psi; t, \tau, a_1\tau, a_2\tau, a_3\tau) \le \Pi(\Phi; t, \tau, a_1\tau, a_2\tau, a_3\tau).$$

Denote $\mathfrak{h}_1 = (a_1, 0, 0), \mathfrak{h}_2 = (0, a_2, 0), \mathfrak{h}_3 = (0, 0, a_3)$. It holds:

$$\Pi(\Psi; t, \tau, a_1\tau, a_2\tau, a_3\tau)(x) - \Pi(\Phi; t, \tau, a_1\tau, a_2\tau, a_3\tau)(x) = \Psi(x) - \Phi(x)$$

$$-\tau \max_{v\in Q}\min_{u\in P} \sum_{m=1}^{3} \Big(\frac{\Psi(x) - \Psi(x - \mathfrak{h}_m\tau)}{a_m\tau} f_m^+ + \frac{\Psi(x + \mathfrak{h}_m\tau) - \Psi(x)}{a_m\tau} f_m^-\Big)$$

$$+\tau \max_{v\in Q}\min_{u\in P} \sum_{m=1}^{3} \Big(\frac{\Phi(x) - \Phi(x - \mathfrak{h}_m\tau)}{a_m\tau} f_m^+ + \frac{\Phi(x + \mathfrak{h}_m\tau) - \Phi(x)}{a_m\tau} f_m^-\Big).$$

By rearranging terms and using the obvious relations $f_m^+ - f_m^- = |f_m|$ and $\max_v \min_u g_1(u, v) - \max_v \min_u g_2(u, v) \le \max_v \max_u \big[g_1(u, v) - g_2(u, v)\big]$, one obtains

$$\Pi(\Psi; t, \tau, a_1\tau, a_2\tau, a_3\tau)(x) - \Pi(\Phi; t, \tau, a_1, a_2, a_3)(x) \le \Psi(x) - \Phi(x)$$

$$+\tau \max_{v\in Q}\max_{u\in P} \sum_{m=1}^{3} \Big[\Big(\frac{\Psi(x - \mathfrak{h}_m\tau) - \Phi(x - \mathfrak{h}_m\tau)}{a_m\tau} - \frac{\Psi(x) - \Phi(x)}{a_m\tau}\Big) f_m^+$$

$$+\Big(\frac{\Psi(x + \mathfrak{h}_m\tau) - \Phi(x + \mathfrak{h}_m\tau)}{a_m\tau} - \frac{\Psi(x) - \Phi(x)}{a_m\tau}\Big)(-f_m^-)\Big]$$

$$\le \Psi(x) - \Phi(x) - \tau \sum_{m=1}^{3} \frac{|f_m|}{a_m\tau}\big(\Psi(x) - \Phi(x)\big) = \Big(1 - \sum_{m=1}^{3} \frac{|f_m|}{a_m}\Big)\big(\Psi(x) - \Phi(x)\big).$$

With $\sum_{m=1}^{3} |f_m| \le \sqrt{3}M$, one arrives at the inequality $1 - \sum_{m=1}^{3} \frac{|f_m|}{a_m} \ge 0$, which implies the required inequality. □

Lemma 3.3. (Generator property). *The generator property holds for the operator* Π.

Proof. Let $\phi \in C_b^2(R^3)$. Denote $\mathfrak{d}_1 = (h_1, 0, 0), \mathfrak{d}_2 = (0, h_2, 0), \mathfrak{d}_3 = (0, 0, h_3)$. We have

$$\Pi(\phi; t, \tau, h_1, h_2, h_3)(x) = \phi(x) -$$

$$\tau \max_{v\in Q}\min_{u\in P} \sum_{m=1}^{3} \Big(\frac{\phi(x) - \phi(x - \mathfrak{d}_m)}{h_m} f_m^+ + \frac{\phi(x + \mathfrak{d}_m) - \phi(x)}{h_m} f_m^-\Big).$$

Estimate

$$\Big|\frac{\Pi(\phi;t,\tau,h_1,h_2,h_3)(x)-\phi(x)}{\tau}+\max_{v\in Q}\min_{u\in P}\langle D\phi(x),f\rangle\Big| =$$

$$\Big|-\max_{v\in Q}\min_{u\in P}\sum_{m=1}^{3}\Big(\frac{\phi(x)-\phi(x-\mathfrak{d}_m)}{h_m}f_m^+ + \frac{\phi(x+\mathfrak{d}_m)-\phi(x)}{h_m}f_m^-\Big)+$$

$$\max_{v\in Q}\min_{u\in P}\sum_{m=1}^{3}\frac{\partial\phi}{\partial x_m}(f_m^+ + f_m^-)\Big|$$

$$\leq \max_{u\in P}\max_{v\in Q}\sum_{m=1}^{3}\Big|\Big(\frac{\partial\phi}{\partial x_m}-\frac{\phi(x)-\phi(x-\mathfrak{d}_m)}{h_m}\Big)f_m^+ + \Big(\frac{\partial\phi}{\partial x_m}-\frac{\phi(x+\mathfrak{d}_m)-\phi(x)}{h_m}\Big)f_m^-\Big|$$

$$\leq M\,\|D^2\phi\|\sum_{m=1}^{3}h_m.$$

Here, M is the bound of $|f|$. Choosing $h_m = a_m\tau$ and letting $C = \sum_{m=1}^{3} a_m$ yields

$$\Big|\frac{\Pi(\phi;t,\tau,h_1,h_2,h_3)(x)-\phi(x)}{\tau}+\mathsf{H}(t,x,D\phi(x))\Big|\leq M\,C\,\|D^2\phi\|\,\tau.$$

□

3.2.4. Three Dimensional Simulation of Cell Shrinkage

The finite difference scheme based on the operator Π is implemented as a parallelized program on a Linux cluster. We compute the evolution of the cell boundary during freezing. The function f in our case is

$$f(t,u,v)=\alpha\big(c_{\mathrm{out}}(t)-c_{\mathrm{in}}(t)\big)^+u+\alpha\big(c_{\mathrm{out}}(t)-c_{\mathrm{in}}(t)\big)^-v,\quad u,v\in R^3,\quad |u|\leq 1,\ |v|\leq 1.$$

Here the controls u and v are responsible for the outflow and inflow, respectively. The spatial grid $200\times 200\times 200$ for the cubic region $0.3\times 0.3\times 0.3$ was utilized, and 1000 time steps of size $\tau = 0.001$ were done. The runtime is about 26 minutes on 30 threads. The initial shape is presented in Figure 40a. Figures 40b and 40c show the computed shape at the time instants $t = 0.8$ and $t = 0.92$.

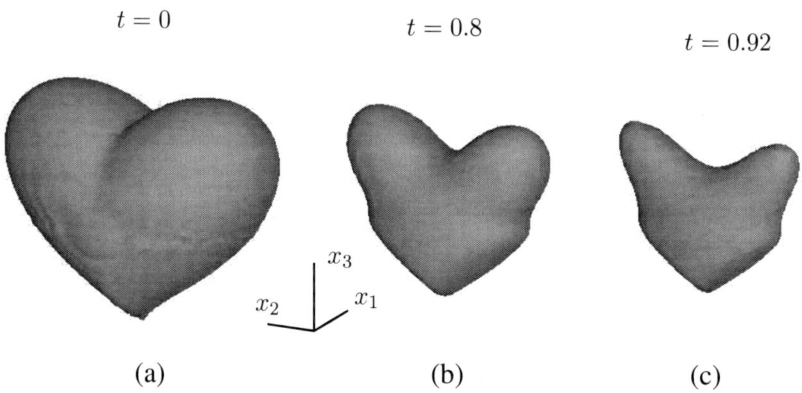

Figure 40. Osmotic cell shrinkage during freezing.

3.2.5. Accounting for the Membrane Tension Using Reachable Set Approach

In reality, the deformation of the cell membrane depends on the membrane tension which is a function of the curvature. Therefore, a more realistic expression for the normal velocity of the cell boundary would be:

$$\mathcal{V}(t,x) = \alpha(c_{\text{out}}(t) - c_{\text{in}}(t)) + \gamma\sigma(x),$$

where $\sigma(x)$ is the angular curvature at the current point x of the cell boundary, and γ is a constant. The angular curvature is explained in Figure 41. The resulting Hamiltonian is

$$\mathsf{H}(t,x,p) = -\big(\alpha(c_{\text{out}}(t) - c_{\text{in}}(t)) + \gamma\sigma(x)\big)|p|. \tag{57}$$

Note that accounting for the curvature can alter the convexity/concavity structure of the Hamiltonian depending on the state x.

Instead of finding viscosity solutions to equation (54), level sets of the value function of an appropriate conflict control problem with the Hamiltonian (57) are computed. The problem is treated within the framework of [12, 31], and [34]. The consideration is carried out in R^2.

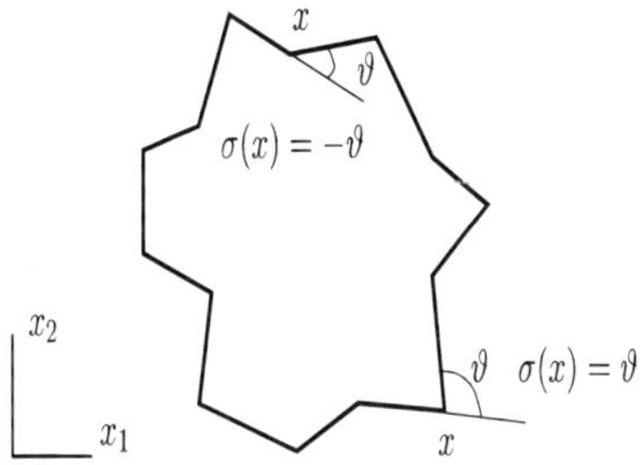

Figure 41. Explanation of the angular curvature.

It is easily seen that the conflict control system of the form

$$\dot{x} = \alpha\big(c_{\text{out}}(t) - c_{\text{in}}(t) + \gamma\sigma(x)\big)^{+} u + \alpha\big(c_{\text{out}}(t) - c_{\text{in}}(t) + \gamma\sigma(x)\big)^{-} v; \quad x, u, v \in R^2,\ |u| \le 1,\ |v| \le 1 \tag{58}$$

has the Hamiltonian (57). The initial cell shape $\Sigma_0 = \Sigma(0)$ is considered as the target set of this differential game where the control u strives to bring the state vector of (58) to Σ_0, and the objective of the control v is opposite.

Since only one of the controls u and v is active at every time t and every point x (see the definition of the operations "+" and "−"), one can reduce the construction of level sets of the value function of system (58) to finding reachable sets of the control system

$$\dot{x} = \text{sign}\mathcal{V} \cdot u\,; \quad x, u \in R^2, \quad u \in |\mathcal{V}|\, S, \quad S = \{x \in R^2 : |x| \le 1\} \tag{59}$$

with $\mathcal{V} = \alpha\big(c_{\text{out}}(t) - c_{\text{in}}(t) + \gamma\sigma(x)\big)$ and the initial set Σ_0.

Denote by $x(t; x_0, u(\cdot))$ the state vector of system (59) at time $t \ge 0$ provided that x_0 is the initial point at time $t = 0$, and $u(\cdot)$ is an admissible measurable control acting on the time interval $[0, t]$. The set of points reachable from Σ_0 at time t is

$$G(t, \Sigma_0) = \bigcup_{x_0 \in \Sigma_0} \bigcup_{u(\cdot)} x(t; x_0, u(\cdot)).$$

The algorithm for the numerical construction of the sequence $\{\mathcal{G}_i = G(i\Delta t, \mathcal{G}_{i-1}),\ i = 1, 2, ..., \mathcal{G}_0 = \Sigma_0\}$, is similar to that described in [41]. The difference is that the cases of local concavity and local convexity alter depending on the sign of $\mathcal{V}$.

Simulation results showing the time evolution of the cell shape during freezing are presented in Figure 42. The sequence of reachable sets is computed on the time interval $[0, 0.645]$ with the time step $\Delta t = 0.001$, every eighth set is drawn. The restriction on the control vector is $u \in |\mathcal{V}|\, P \cdot S$, where $P = \{p_{ij}\}$ is the 2×2- diagonal matrix accounting for anisotropy ($p_{11} = 3$, $p_{22} = 1$). For the left picture, $\gamma = 0$, i.e. the curvature independent case is considered.

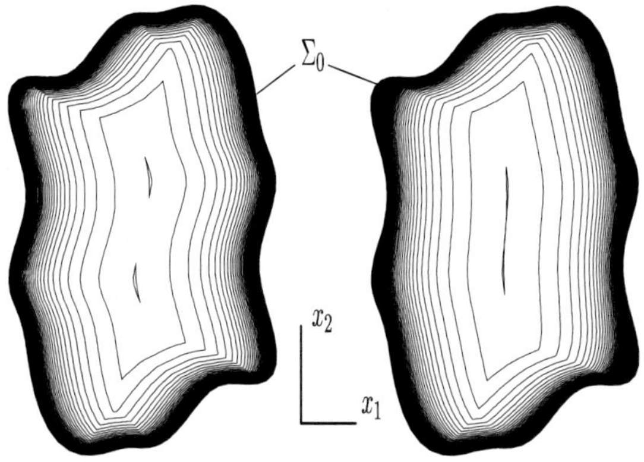

Figure 42. Time evolution of the cell shape during freezing. To the left: without accounting for the curvature. To the right: with accounting for the curvature.

For the right picture, $\gamma = 0.06$. Note that the low propagation velocity at the beginning of the process results in the accumulation of lines (dark regions) near the initial cell boundary.

In Figure 43, an example of the simulation of cell rehydration during thawing is presented. The computation is done on the time interval $[0, 9.7]$ with the time step $\Delta t = 0.002$. Every 200th reachable set is drawn. At the end of the process, the stabilization of the sets is observed, because the equilibrium between the salt concentrations inside and outside the cell is achieved.

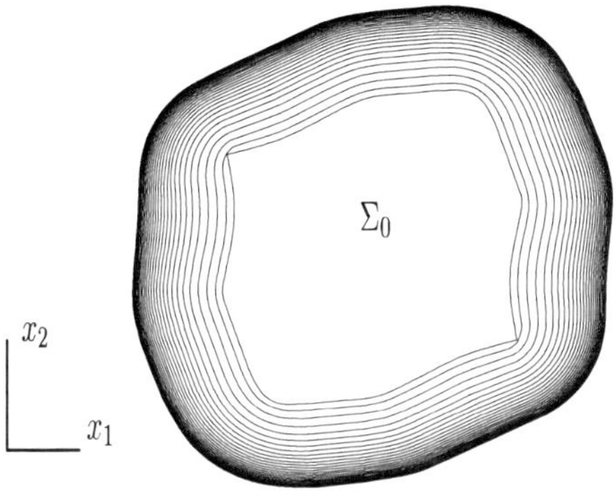

Figure 43. Time evolution of the cell shape during thawing with accounting for the curvature.

Conclusion

This chapter was concentrated on the application of modern mathematical methods of optimal control to problems related to freezing and thawing of living cells. The aim of the investigation was to show that not only cooling and thawing rates are important but also changes of regimes that ensure simultaneous freezing (thawing) of intra- and extracellular liquid (ice). Such complex protocols allow to reduce large stresses exerted on cell membranes and osmotic flows causing shrinkage or swelling of cells. Another advance of optimal control techniques is that they support the design of cooling protocols ensuring a smooth transition from the regime of supercooling to the frozen state, which is also favorable for the viability and the fresh level of frozen-thawed cells. Thus, our, mostly mathematical, study outlines new approaches to the improvement of conventional cooling-freezing protocols. We are interested in drawing attention of experimenters to the validation of our approach.

Acknowledgment

The authors appreciate the support of the German Research Society (Deutsche Forschungsgemeinschaft) in the framework of the intention "Optimization with partial differential equations" (SPP 1253).

References

[1] Toner, M., Cravalho, E.G., & Karel, M. (1990). Thermodynamics and kinetics of intracellular ice formation during freezing of biological cells. *Journal of Applied Physics*, February 1, 67(3), 1582–1593.

[2] Karlsson, J. O. M., Cravalho, E. G., Borel Rinkes, I. H., Tompkins, R. G., Yarmush, M. L., & Toner, M. (1993). Nucleation and growth of ice crystals inside cultured hepato-cytes during freezing in the presence of dimethyl sulfoxide. *Biophysical J.*, December, 65(6), 2524–2536.

[3] Kiselev, M. A., Lesieur, P., Kisselev, A. M. & Ollivon, M. (2000). Ice formation in model biological membranes in the presence of cryoprotectors. *Nuclear Instruments & Methods in Physics Research* A, 448, 225–260.

[4] Xu, Y., McDonough, J. M., Tagavi, K. A., & Gao, D. (2004). Two-dimensional phase-field model applied to freezing into supercooled melt. *Cell Preservation Technology*, June, 2(2), pp. 113–124.

[5] Xu, Y., Yang, T., McDonough, J. M., & Tagavi, K. A. (2003). Parallelization of phase-field model for phase transformation problem. In B. Chetverushkin, J. Periaux, A. Ecer, & J. Satofuka (Eds.), Parallel Computational Fluid Dynamics 2003: *Advanced Numerical Methods, Software and Applications* (pp. 213–218). Amsterdam, North-Holland: Elsevier.

[6] Mazur, P. (2004). Principles of cryobiology. In B. J. Fuller, N. Lane, & E. E. Benson (Eds.), *Life in the Frozen State* (pp. 3–65). Boca Raton, FL: CRC Press.

[7] Muldrew, K., Acker, J. P., Elliott, J. A. W., & McGann, L. E. (2004). The water to ice transition: Implications for living cells. In B. J. Fuller, N. Lane, & E. E. Benson (Eds.), *Life in the Frozen State* (pp. 67–108). Boca Raton, FL: CRC Press.

[8] Irimia, D., & Karlsson, J. O. M. (2005). Kinetics of intracellular ice formation in one-dimensional arrays of interacting biological cells. *Biophysical Journal*, 88, 647–660.

[9] Mukherjee, I N. A Rational Design Approach for the Cryopreservation of Natural and Engineered Tissues. *PhD Thesis.* Atlanta, GA: Georgia Institute of Technology; 2008.

[10] Caginalp, G. (1986). An analysis of a phase field model of a free boundary. *Arch Rat. Mech. Anal.*, 92, 205–245.

[11] Hoffmann, K.-H.,& Lishang, J. (1992). Optimal control of a phase field model for solidification. *Numer. Funct. Anal. Optimiz.*, 13 (1&2), 11–27.

[12] Krasovskii, N. N., & Subbotin, A. I. (1988). *Game-Theoretical Control Problems.* New York: Springer.

[13] Crandall, M. G., & Lions, P. L. (1983). Viscosity solutions of Hamilton - Jacobi equations. *Trans. Amer. Math. Soc.*, 277, 1–47.

[14] Caginalp, G., & Chen, X. (1998). Convergence of the phase field model to its sharp interface limits. *European Journal of Applied Mathematics*, 9, 417–445.

[15] Lions, J. L. (1971). Optimal Control of Systems Governed by Partial Differential Equations. *Grundlehren der mathematischen Wissenschaften* 170. Berlin, New York: Springer-Verlag.

[16] Hoffmann, K.-H., Amler, T. G., Botkin, N. D., & Ruf, K. (2012). Regularity of solutions of a phase field model [Preprint-Nr.: SPP1253-141]. http://www.am.uni-erlangen.de/home/spp1253/wiki/images/b/b6/Preprint-SPP1253-141.pdf

[17] Amler, T. G., Botkin, N. D., Hoffmann, K.-H., & Hoteit, I. (2012). Continuity in time of solutions of a phase-field model [Preprint-Nr.: SPP1253-142]. http://www.am.uni-erlangen.de/home/spp1253/wiki/images/7/77/Preprint-SPP1253-142.pdf

[18] Fletcher, R., & Reeves, C. M. (1964). Function minimization by conjugate gradients. *Computer Journal*, 7, 149–154.

[19] Lions, J. L., & Magenes, E. (1972). Non-Homogeneous Boundary Value Problems and Applications: Vol. I. *Grundlehren der mathematischen Wissenschaften* 181. Berlin, New York: Springer-Verlag.

[20] Hoffmann, K-H., & Botkin, N. D. (2008). Optimal control in cryopreservation of cells and tissues. *Advances in Mathematical Sciences and Applications*, 29, 177–200.

[21] Frémond, M. (2002). *Non-Smooth Thermomechanics*. Berlin: Springer-Verlag.

[22] Haeupl P., & Xu, Y. (2001). Numerical simulation of freezing melting in porous materials under the consideration of the coupled heat and moisture transport. *Journal of Thermal Envelope and Bilding Science*, 25(1), 4–31.

[23] Nguetseng, G. (1989). A general convergence result for a functional related to the theory of homogenization. *SIAM J. Math. Anal.*, 20(3), 608–623.

[24] Allaire, G. (1992). Homogenization and two-scale Convergence. *SIAM J. Math. Anal.*, 23(6), 1482–1518.

[25] Jikov, V. V., Kozlov, S. M., & Oleinik, O. A. (1994). *Homogenization of Differential Operators and Integral Functionals*. Berlin: Springer-Verlag.

[26] Hoffmann, K.-H., & Botkin, N. D. (2000). Homogenization of von Kármán plates excited by piezoelectric patches. ZAMM Z. *Angew. Math. Mech.*, 80(9), 579–590.

[27] Hoffmann, K.-H., Botkin, N. D., & Starovoitov, V. N. (2005). Homogenization of interfaces between rapidly oscillating fine elastic structures and fluids. *SIAM J. Appl. Math.*, 65(3), 983–1005.

[28] Batycky, R. P., Hammerstedt, R., & Edwards D. A. (1997). Osmotically driven intracellular transport phenomena. *Phil. Trans. R. Soc. Lond.* A, 355, 2459–2488.

[29] Mao, L., Udaykumar, H. S., & Karlsson J. O. M. (2003). Simulation of micro scale interaction between ice and biological cells. *Int. J. of Heat and Mass Transfer*, 46, 5123–5136.

[30] Chen, S. C., Mrksich, M., Huang, S., Whitesides, G. M., & Ingber, D. E. (1997). Geometric control of cell life and death. *Science*, 276, 1425–1428.

[31] Krasovskii, N. N, & Subbotin, A. I. (1974). *Positional Differential Games*. Moscow: Nauka (in Russian).

[32] Tchir, J., & Acker, J. (2010). Mitochondria and membrane cryoinjury in micropatterned cells: Effects of cell-cell interactions. *Cryobiology*, 61, 100–107.

[33] Subbotin, A. I., & Chentsov, A. G. (1981). *Optimization of Guaranteed Result in Control Problems*. Moscow: Nauka (in Russian).

[34] Subbotin, A. I. (1995). *Generalized Solutions of First Order PDEs*. Boston: Birkhäuser.

[35] Botkin, N. D., Hoffmann, K.-H., Mayer, N., Turova, V. L. (2011). Approximation schemes for solving disturbed control problems with non-terminal time and state constraints. *Analysis*, 31, 355–379

[36] Souganidis, P. E. (1985). Approximation schemes for viscosity solutions of Hamilton-Jacobi equations. *J. Differ. Equ.*, 59, 1–43.

[37] Malafeyev, O. A., & Troeva, M. S. (1998). A weak solution of Hamilton-Jacobi equation for a differential two-person zero-sum game. *Preprints of the Eight Int. Symp. on Differential Games and Applications*, Maastricht, Netherland, July 5-7, 1998, 366–369.

[38] Botkin, N. D., Hoffmann, K-H., & Turova, V. L. (2011). Stable numerical schemes for solving Hamilton-Jacobi-Bellman-Isaacs equations. *SIAM Journal on Scientific Computing*, 33 (2), 992–1007.

[39] Botkin, N. D., Hoffmann, K-H., & Turova, V. L. (2011). Optimal control of ice formation in living cells during freezing. *Applied Mathematical Modelling*, 35, 4044–4057.

[40] Karlsson, J.O.M. (2001). A theoretical model of intracellular devitrification. *Cryobiology*, 42, 154–169.

[41] Patsko, V. S., & Turova, V. L. (2009). From Dubins' car to Reeds and Shepp's mobile robot. *Comput. Visual. Sci.*, 13(7), 345–364.

INDEX

A

acid, 22
activation energy, 78, 161
additives, 22, 97
adipose tissue, 79, 90
adult stem cells, 79, 90
adverse conditions, 10, 24
agar, viii, 1, 6, 7, 8, 9, 11, 12, 13, 14, 17, 18, 19, 30, 35, 37, 38, 39, 40, 42, 43, 50, 52
aggregation, 111, 160
agriculture, 3, 26, 33, 50
alcohols, 22, 87
algae, 17, 35
aluminium, 69
amino, 13, 22, 87
amino acids, 22
ammonium, 22
anisotropy, 169
antibiotic, 4, 72
apoptosis, 98
apoptotic pathways, 102
aqueous solutions, 87, 88
arbuscular mycorrhizal fungi, 13, 64
Artemia, 97
assessment, 4, 27, 88
Austria, 113, 114

B

background radiation, 18
bacteria, 6, 14, 17, 19, 23, 30, 44, 49, 56, 75, 97
banking, 92, 93, 100, 103
banking industry, 100
banks, ix, 91, 92
base, 98, 99, 101, 114, 164
benefits, 103
biochemistry, 21, 48
biodegradation, 38
biodiversity, 2, 3
biological activity, vii
biological control, 16, 31, 34, 36
biological systems, 48
biomass, 4, 39, 63
biomolecules, 58
biopolymer, 95
biopreservation, 106
biotechnological applications, 28
biotechnology, 21, 25, 28, 59, 60
blood, 22, 92, 95, 98, 100, 103, 105
Boltzmann constant, 161
bone, 92, 99, 106, 107
bone marrow, 92, 106, 107
bounds, 150
Brazil, 5, 17, 63
breeding, 6, 21, 25, 50

C

calcium, 26, 99, 107
calorimetry, vii, 68, 88
carbonic acids, 22
CBS, 9, 10, 14, 20, 25
cell biology, 100
cell culture, 98
cell death, vii, 99
cell differentiation, 93
cell division, 20

cell lines, 35, 102, 103
cell membranes, ix, 15, 24, 57, 109, 141, 143, 171
cell size, 20
central bank, 29
ceramic, viii, 2, 16, 17, 55
cheese, 36
chemical, ix, 21, 22, 24, 26, 33, 48, 109, 110
chemical reactions, 110
chemicals, 21
chlorophyll, 29, 30
CHO cells, 96, 97
chromatography, 4
chromosomal abnormalities, 27
classification, 35
clinical trials, 92
colonization, 26, 42, 66
combined effect, 111
commercial, 3, 20, 24, 44, 55
community, 101
compensation, 68, 69, 70
competition, 26
complexity, ix, 109, 112
composition, 20, 21, 24, 25, 56, 98, 107
compounds, 21, 22, 87
computation, 149, 151, 157, 169
computer, 113, 114, 157
computing, ix, 110, 117
conductivity, 125, 161
conflict, 142, 164, 168
congelation, viii, 67
Congress, 49, 51, 59, 64
conservation, ix, 42, 45, 46, 54, 66, 109, 162, 163
constant rate, 69, 71, 84
construction, 19, 169
contact time, 38
contamination, vii, viii, 1, 2, 3, 8, 18, 30, 103
control group, 44
convergence, 146, 147, 148, 165, 173
cooling, vii, viii, ix, 13, 15, 20, 23, 24, 25, 26, 30, 33, 36, 54, 58, 67, 69, 70, 71, 73, 74, 75, 76, 77, 78, 79, 82, 84, 85, 86, 87, 88, 90, 103, 109, 110, 111, 112, 113, 114, 115, 118, 120, 121, 125, 133, 134, 135, 141, 142, 143, 145, 150, 151, 152, 153, 155, 156, 170, 171
correlation, 10, 79
corticosteroid therapy, 102
cost, 3, 6, 148
cotton, 53
Croatia, 49
crops, 31, 41
crown, 30
cryopreservation, vii, viii, ix, 1, 2, 3, 4, 5, 6, 7, 8, 9, 13, 14, 15, 16, 17, 18, 19, 20, 21, 22, 23, 24, 25, 26, 27, 28, 29, 30, 31, 33, 34, 35, 36, 37, 38, 39, 40, 41, 42, 43, 45, 46, 47, 48, 52, 55, 56, 57, 58, 59, 60, 62, 63, 64, 65, 66, 67, 72, 73, 78, 86, 88, 89, 90, 91, 92, 93, 94, 98, 100, 101, 102, 103, 107, 112, 113, 173
cryoprotectant, vii, 1, 18, 20, 21, 22, 25, 37, 38, 39, 41, 43, 44, 66
cryosurgery, 79
crystal growth, 73, 75
crystalline, 13, 15
crystallisation, viii, ix, 23, 67, 68, 70, 71, 72, 73, 74, 75, 78, 79, 80, 82, 83, 84, 85, 86
crystallization, vii, 13, 14
crystals, ix, 14, 20, 22, 23, 83, 109, 110, 111, 114, 141, 160, 171
cultivation, 3, 9, 24, 41, 46, 51
culture, ix, 2, 4, 6, 7, 8, 9, 10, 12, 14, 15, 18, 24, 27, 30, 31, 35, 36, 37, 39, 41, 42, 43, 45, 50, 51, 53, 55, 60, 63, 73, 91, 96, 98, 99, 100, 108
culture media, 43
culture medium, 43, 99
cycles, 19, 32, 36, 40
cytoplasm, 21, 111
Czech Republic, 1

D

damages, 82, 111
death rate, 24
decay, 12, 52
deconvolution, 79, 85
deficiencies, 105
deformation, 168
degenerate, 9, 15
degradation, 2
dehydrate, 20

dehydration, ix, 13, 14, 20, 23, 25, 26, 42, 56, 61, 74, 79, 80, 84, 85, 94, 106, 109, 110, 111, 112, 141, 142, 161
denaturation, 13, 94
dendrites, 111, 116, 141, 160, 161
derivatives, 22, 129, 145, 147, 148, 164
desiccation, ix, 91, 92, 94, 96, 97, 99, 100, 101, 104, 106
desorption, 94
detection, 85
deviation, 125, 127, 129
differential scanning, vii, viii, 67, 68, 82, 88, 89, 90
differential scanning calorimeter, viii, 67, 88
differential scanning calorimetry, viii, 67, 89, 90
diffusion, 73, 74, 78, 89
diffusivity, 71
dimethylformamide, 87
dimorphism, 27
discretization, 147, 164
discs, 9, 11, 17
diseases, 27, 35, 36
displacement, 136, 138
distilled water, 9, 10, 11, 12, 15, 35, 38, 49, 50, 51, 52, 72, 84
distortions, 26
distribution, 6, 18, 36, 40, 48, 131, 133, 134, 135, 136, 138, 142, 162
DNA, 5, 17, 30, 32, 33, 38, 44
DOI, 108
donors, 92
drugs, 34
drying, vii, 1, 3, 7, 8, 9, 10, 12, 13, 14, 15, 26, 28, 32, 36, 37, 42, 45, 46, 48, 49, 50, 52, 53, 54, 55, 56, 57, 60, 63, 64, 94, 95, 96, 97, 104, 105, 106
DSC, 68, 69, 70, 71, 72, 73, 74, 78, 79, 80, 81, 82, 83, 85, 86, 87, 90

E

ECM, 43
edible mushroom, 5, 43
elaboration, viii, 68, 81
electrolyte, 23
electron microscopy, 26
electroporation, 33, 34, 95, 104
embryogenesis, 97, 106
embryonic stem cells, 95, 99, 102, 103, 107
encapsulation, 7, 25, 26, 29, 42, 61, 65
endothelial cells, 99, 106
endothermic, 68, 71
energy, 125, 132, 144, 150, 154, 157
energy density, 132
England, 28
environment, 21, 26
environmental conditions, 7, 37
environmental control, 93
environmental factors, 36
enzymatic activity, 39
enzymes, 3, 4, 15, 38, 39, 40, 45, 46, 54
epithelial cells, 96, 99, 105, 107
equilibrium, 20, 80, 88, 132, 169
equipment, 7, 9, 12, 18, 24, 98
erythrocytes, 95, 105
ethanol, 13
ethyl alcohol, 22
ethylene, 13, 22, 25, 57, 68, 72, 88, 89
ethylene glycol, 13, 22, 25, 68, 72, 88, 89
eukaryotic, 21, 104
eukaryotic cell, 21, 104
evolution, 107, 142, 167, 169, 170
experimental condition, 112
exposure, 20, 36, 41, 93, 98
extracellular matrix, 114, 131, 141, 142, 143
extracts, 75, 82, 84

F

fibroblasts, 85, 97
fingerprints, 27, 32
fish, 23, 78, 90
fixation, 26
flexibility, 106
fluctuations, 93
fluid, 15, 22, 35, 57, 94, 111, 143, 144, 150, 154, 160, 161, 163
fluorescence, 30
food industry, 36
formamide, 22
formation, viii, ix, 14, 23, 24, 28, 60, 67, 68, 72, 79, 82, 83, 84, 85, 86, 89, 90, 98, 105, 109, 111, 112, 114, 131, 142, 157, 161, 171, 172, 175
formula, 122, 126, 130, 132, 144, 147
France, 60, 67

free energy, 123, 132
freezing, vii, viii, ix, 1, 6, 7, 8, 9, 10, 13, 14, 17, 19, 20, 21, 23, 24, 25, 26, 30, 31, 32, 36, 37, 38, 39, 40, 42, 43, 44, 45, 46, 48, 49, 50, 54, 56, 57, 58, 59, 60, 63, 64, 65, 66, 67, 68, 72, 73, 74, 78, 79, 80, 82, 83, 85, 86, 87, 88, 89, 90, 91, 92, 93, 94, 96, 97, 98, 103, 109, 110, 111, 112, 113, 114, 115, 123, 131, 132, 138, 140, 141, 142, 143, 144, 146, 150, 151, 158, 159, 160, 161, 167, 169, 170, 171, 173, 175
frost, 21
frost resistance, 21
fructose, 16
funds, 9, 12, 24
fungi, vii, viii, 1, 2, 3, 4, 5, 6, 7, 8, 9, 10, 11, 12, 13, 14, 15, 16, 17, 18, 20, 21, 22, 23, 24, 25, 26, 27, 28, 29, 30, 31, 32, 33, 34, 35, 36, 37, 38, 40, 41, 42, 43, 44, 45, 46, 47, 48, 49, 50, 51, 52, 53, 54, 55, 56, 57, 58, 59, 60, 61, 63, 64, 65, 66, 94
fungus, 5, 7, 11, 14, 15, 19, 22, 25, 26, 27, 28, 29, 30, 31, 32, 33, 35, 36, 42, 47, 49, 53, 54, 55, 57, 59, 61, 62, 63, 65
fusion, 76

G

gastrointestinal tract, 15, 34, 63
gel, 7, 9, 12, 13, 15, 30, 31, 32, 34, 35, 51, 53, 59, 62
genomic stability, 47
genus, 5, 12, 21, 38, 57
Georgia, 172
Germany, 109
germination, 12, 13, 42
Gigaspora margarita, 41
glass transition, 20, 57
glass transition temperature, 20, 57
glasses, 87
Glomus intraradices, 28, 41
glucose, 14, 21, 22, 39, 95
glutamic acid, 22
glutathione, 98
glycerol, 13, 17, 21, 22, 25, 26, 30, 31, 33, 36, 37, 38, 39, 41, 42, 43, 44, 57, 72, 73, 81, 87, 88, 89, 96, 112
glycine, 22
glycol, 13, 17, 21, 22, 39, 57, 81, 87, 90
glycoproteins, 22
green alga, 29, 62
grouping, 25
growth, viii, ix, 2, 3, 4, 5, 7, 8, 9, 10, 12, 14, 15, 16, 19, 20, 22, 24, 26, 29, 30, 32, 33, 34, 35, 38, 39, 40, 41, 43, 45, 47, 49, 63, 64, 74, 89, 97, 98, 99, 100, 109, 110, 111, 112, 114, 152, 157, 158, 160, 171
growth rate, 4, 9, 10, 15, 16, 38, 39, 64
guidelines, 3, 20

H

habitats, 35
Hamiltonian, 122, 146, 149, 163, 164, 168, 169
Hamilton-Jacobi equations, ix, 174
harvesting, 92
health, 2, 97
heat conductivity, 116, 125, 144
heat release, 74, 75, 76, 79, 80, 85, 116, 120, 133
heat transfer, 116, 117, 144
helium, 69
hepatocytes, 92, 99, 102, 106, 107, 112
hepatoma, 106
homogeneity, 5, 17
horses, 79
host, 13, 27, 29, 31, 32, 33, 66, 102
human, 10, 29, 78, 85, 89, 92, 95, 96, 97, 99, 102, 103, 104, 105, 106, 107, 108
humidity, 9
Hunter, 105
hydrogels, 99
hydrogen peroxide, 4, 40
hydroxyl, 87
hypothermia, 98

I

images, 172, 173
imbalances, 96, 98
immersion, 23, 43, 44
immune system, 96
impregnation, 86
improvements, 27, 92, 94, 97
in vitro, 13, 15, 33, 41, 42, 47, 52, 63, 73, 88

in vivo, 15, 33, 41
incubator, 36
induction, 79
industry, ix, 3, 46, 91, 92, 97, 98, 102
inequality, 166
infection, 26, 27, 32, 33, 34
infestations, 10
inhibition, 22
inhibitor, 31, 99
initiation, 36, 98
injury, 23, 24, 48, 58, 79, 95, 96, 98, 105, 106
inoculum, 34, 41, 65
inositol, 15
insects, 21
insulin, 113
integration, 128
integrity, 3, 4, 5, 27, 35, 46, 96, 97, 104
interface, 112, 121, 123, 124, 172
invertebrates, 23
iron, 98, 99
isolation, 33, 66

J

Japan, 6
Jordan, 107

K

karyotype, 99
kinetics, 111, 113, 171

L

lactic acid, 19
lactose, 21, 22
larvae, 15, 16, 34, 54, 104
lead, 21, 79, 98
leaks, 18
lichen, 29, 62
life cycle, 41
light, 26, 29, 58, 59
lignin, 38
liquid chromatography, 4
liquid phase, 14
liquids, 138, 140, 146
logistics, 100
longevity, 52
long-term maintenance, vii, 1, 3, 8, 18, 40, 45
low temperatures, vii, 6, 16, 19, 24, 25, 36, 46, 48, 54, 55, 56, 93
lysis, 74, 75

M

macromolecules, 23
magnitude, 116, 143
malt extract, 22, 39
maltose, 22
mammal, 78
mammalian cells, vii, ix, 91, 94, 95, 98, 100, 101, 104, 105, 106, 107
manganese, 4, 38
mannitol, 21, 22, 41, 42
Markov chain, 112
Marx, 11, 49
mass, 19, 53, 71, 162, 163
materials, 3, 7, 18
mathematical methods, 81, 82
matrix, 26, 131, 147, 169
measurements, 69, 71, 78, 85, 89, 117, 119
mechanical stress, 114
media, 12, 14, 38, 41, 85, 90, 99, 100, 131, 142
medical, ix, 3, 17, 109
medicine, 21, 102, 107
medium composition, 20, 99
melting, viii, 67, 68, 70, 71, 81, 82, 173
melting temperature, viii, 67, 68, 70, 71, 82
membrane permeability, 74, 78
membranes, 23, 58, 171
mesenchymal stem cells, 102, 106, 107
metabolism, 7, 9, 12, 20, 98
metabolites, viii, 2, 3, 4, 16, 28, 45
methanol, 13, 22, 33, 90
methodology, 27, 62
microbial cells, 21
microbial communities, 45
microinjection, 94
micromycetes, 44, 49, 66
microorganisms, 2, 3, 6, 8, 16, 18, 19, 20, 22, 26, 29, 45, 46, 47, 48, 55, 56, 57, 58, 64, 93
microscope, 26, 27

microscopy, 21, 29, 58, 59, 113
mildew, 27, 32
Missouri, 49
models, ix, 49, 77, 83, 109, 111, 113, 114, 123, 141, 142, 151
moderators, 21
modifications, 9, 126
moisture, 12, 36, 61, 94, 95, 96, 173
moisture content, 61, 95, 96
molar volume, 76
molds, 34, 35, 53, 54
mole, 112
molecular weight, 21
molecules, 73, 74
monolayer, 99
morphology, viii, 2, 3, 4, 14, 27, 38, 41, 59, 96
mosaic, 15
Moscow, 174
mutant, 28, 94
mutation, 20
mutations, 27
mycelium, vii, viii, 1, 2, 4, 5, 7, 8, 11, 15, 16, 17, 18, 19, 24, 25, 31, 34, 37, 38, 39, 41, 42, 43, 54
mycorrhiza, 2, 65

N

NaCl, 81, 82, 89
nematode, 34
Netherlands, 50, 51, 59, 64
neurogenesis, 102
neurons, 99
nitrogen, vii, viii, 1, 3, 4, 7, 8, 13, 15, 16, 17, 18, 24, 25, 27, 30, 38, 41, 42, 43, 45, 46, 47, 55, 56, 57, 58, 60, 61, 62, 64, 66, 69, 73, 93, 98, 103, 114
nitrogen gas, 18
nuclear genome, 105
nuclear magnetic resonance, 131, 162
nucleating agent, 75
nucleation, 36, 63, 72, 73, 75, 89, 111, 112, 131, 161
nuclei, 75, 111
nutrient, 2, 24, 40, 51, 98
nutrient media, 40, 51

O

oil, 6, 7, 9, 10, 31, 37, 49, 51
oocyte, 83
operations, 169
optimization, ix, 15, 110, 121, 141
orchid, 29, 42, 61
ordinary differential equations, 144, 145
organ, ix, 98, 100, 109
organism, 5, 11, 18, 23, 27
organs, 101
oscillation, 139
osmolality, 101
osmotic pressure, 80, 142, 161
osmotic stress, 97
ovaries, 86
overproduction, 11
oxygen, 14, 36

P

parallelization, 157
parasites, 6, 28, 32, 34, 52
partial differential equations, 114, 171, 174
pathogens, 10, 11, 29, 31, 36, 47, 62, 63
pathology, 6, 62
penicillin, 28
peptides, 22, 99
perfusion, 38
periodicity, 137
permeability, 76, 77, 87, 88, 89, 161
personal communication, 10, 20
pharmaceutical, 97, 98, 100
phase diagram, viii, 68, 80, 81, 82, 89, 90
phase transformation, 171
phase transitions, 13, 69, 71, 113, 142
phenotypes, 28, 60
physics, 112
Physiological, 57
physiological changes, vii, 1, 8, 28, 45
physiology, 4, 28
plant growth, 41, 43
plants, 21, 23, 41, 56, 59, 113
plasma membrane, 56, 58, 65, 74, 77, 83, 89, 94, 110
plasmid, 33
platelets, 95, 104, 105
ploidy, 59

polycarbonate, 18, 19, 43
polymerase chain reaction, 4, 5, 27, 28, 44, 60
polymorphisms, 5, 27, 28
polypropylene, 18, 19, 22, 40
polystyrene, viii, 2, 16, 17, 19
polyvinyl alcohol, 22
porous materials, 173
porous media, 131
Portugal, 58
potato, 30, 35, 59
precedent, 97
precipitation, 23
preparation, 13, 26
preservation, vii, viii, ix, 1, 2, 3, 4, 5, 6, 7, 8, 9, 11, 12, 13, 14, 15, 16, 17, 18, 20, 23, 24, 25, 26, 27, 28, 29, 30, 31, 32, 33, 34, 35, 36, 37, 38, 39, 41, 44, 45, 46, 47, 48, 49, 50, 51, 52, 53, 54, 55, 56, 58, 59, 60, 61, 62, 63, 64, 65, 75, 91, 93, 94, 95, 98, 99, 100, 102, 103, 105, 107
prevention, 7, 58
production technology, 50
progenitor cells, 103, 105
programming, ix, 110, 142
proliferation, 41
propagation, 24, 41, 42, 65, 163, 169
propane, 13, 25, 33
proposition, 126, 129
propylene, 22, 81, 87, 90
protection, 8, 41, 42, 94, 96
proteins, 22, 57, 94, 97, 98, 103, 106, 107
purity, 6, 11, 52

Q

quality assurance, 15
quality of life, 51, 59, 64
Queensland, 35, 51

R

reactions, vii, 93, 103
reactive oxygen, 98
recovery, ix, 16, 19, 20, 23, 26, 32, 38, 39, 42, 44, 91, 92, 93, 96, 99, 102, 103
recrystallization, 22, 23
red blood cells, 99, 104, 105
regeneration, 54
rehydration, 7, 94, 95, 96, 97, 106, 110, 141, 161, 169
reliable source of cultures, vii
requirements, 3, 5, 13
researchers, 13, 73, 98
resistance, 6, 24, 31, 33, 62
resources, 3, 10, 46, 48, 51
response, 4, 25, 26, 57, 58, 88, 90, 98, 106, 114, 116, 117, 118, 120, 121, 123, 124, 127, 130, 131, 133, 134, 135
risks, vii, 23, 93, 103
room temperature, 10, 12, 14, 29, 30, 31, 35, 40, 42, 93, 94
roots, 13, 26, 27, 42, 59

S

safety, 17, 92, 95
salt concentration, 142, 159, 161, 162, 163, 169
salts, 24, 81
scaling, 132
scanning calorimetry, viii, 67, 68, 88
scanning electron microscopy, 59
sediments, 44
seed, 42, 61, 65, 66
seeding, 74
semen, 21
sensitivity, 24, 28, 34, 64
sensors, 69, 114, 115
serum, 15, 22, 68, 73, 98, 103
serum albumin, 22, 68, 73
shape, ix, 71, 86, 110, 125, 142, 161, 163, 167, 169, 170
shock, 23, 94, 110, 116
showing, 23, 76, 77, 80, 95, 169
silica, 7, 9, 12, 13, 30, 31, 32, 34, 35, 51, 53, 62
simulations, vii, ix, 110, 122, 123, 124, 139, 150, 151, 157, 160, 161, 169, 173
skeletal muscle, 99
slow-freezing protocols, viii, 67, 68, 80, 82, 86
sodium, 88, 90
solid matrix, 131, 132, 136, 137
solidification, 71, 123, 144, 172
solubility, 21

solution, 14, 20, 21, 23, 25, 30, 43, 58, 70, 71, 72, 73, 74, 80, 82, 86, 98, 99, 100, 107, 110, 125, 126, 127, 128, 129, 141, 146, 147, 159, 161, 165, 174
somatic cell, 105
Southern blot, 32
Spain, 32
species, vii, viii, 1, 2, 3, 4, 5, 8, 10, 11, 12, 13, 14, 16, 18, 20, 28, 30, 32, 34, 35, 37, 38, 40, 41, 42, 43, 45, 52, 53, 59, 60, 62, 78, 90, 98
specific heat, 132, 144
spectrophotometric method, 4, 40
sperm, 72, 87, 88, 89, 90
spore, 11, 12, 14, 18, 24, 32, 35, 42, 52
stability, viii, 2, 4, 5, 6, 7, 12, 15, 17, 27, 28, 31, 45, 46, 47, 50, 59, 86, 87, 90
stabilization, 22, 105, 169
starch, 21, 22
state, 2, 6, 9, 16, 20, 25, 30, 37, 56, 57, 84, 86, 89, 111, 112, 118, 120, 125, 126, 133, 136, 146, 148, 149, 157, 159, 161, 168, 169, 171, 174
states, 7, 126, 147
stem cell lines, 103
stem cells, ix, 92, 96, 103, 109
sterile, 7, 9, 10, 11, 12, 15, 17, 35, 38, 39, 49, 50, 51, 52
storage, vii, ix, 1, 2, 3, 4, 5, 6, 7, 8, 9, 10, 11, 12, 13, 14, 15, 17, 18, 19, 20, 21, 22, 23, 25, 28, 29, 30, 31, 32, 33, 34, 35, 36, 37, 38, 39, 40, 41, 42, 43, 45, 46, 47, 49, 50, 51, 52, 53, 54, 55, 56, 57, 58, 60, 61, 62, 63, 64, 65, 66, 91, 92, 94, 95, 96, 98, 99, 100, 101, 102, 103, 105, 106, 107, 112
storage media, 98
stress, 19, 97, 98, 106, 136, 138, 141, 143
striatum, 107
stroke, 102
structural modifications, 87
structure, 15, 112, 136, 138, 139, 149, 161, 168
substrates, 9, 12, 13, 17, 22, 27, 29
success rate, 95
sucrose, 14, 21, 25, 32, 41, 42, 89
sugar beet, 31, 32
supercooling, ix, 109, 110, 112, 116, 131, 133, 150, 171
surface area, 76, 117
surface tension, 112
surfactants, 22
survival, viii, ix, 2, 3, 4, 10, 11, 18, 19, 20, 21, 22, 23, 24, 25, 27, 29, 30, 31, 33, 36, 38, 39, 40, 42, 43, 44, 47, 63, 65, 73, 84, 85, 86, 90, 92, 99, 104, 107, 109, 111, 142
survival rate, ix, 10, 11, 25, 73, 85, 109, 111, 142
suspensions, 9, 11, 18, 19, 24, 35, 74, 82, 85, 86, 88, 89
swelling, 110, 141, 142, 163, 171
symbiosis, 2, 30, 41
synchronization, 36
synergistic effect, 97

T

tanks, 20
taxa, 6, 35
techniques, viii, ix, 2, 3, 5, 6, 7, 9, 11, 17, 25, 26, 27, 28, 32, 35, 46, 48, 49, 59, 60, 62, 104, 110, 113, 121, 122, 131, 136, 137, 142, 161, 171
technologies, 12, 52
technology, 25, 26, 51
temperature, viii, ix, 7, 12, 14, 17, 20, 21, 22, 24, 25, 31, 36, 42, 56, 57, 58, 61, 62, 63, 64, 67, 68, 69, 70, 71, 75, 76, 77, 80, 82, 83, 86, 89, 93, 94, 106, 107, 109, 110, 111, 112, 114, 115, 116, 117, 118, 119, 120, 121, 123, 124, 125, 127, 129, 130, 131, 132, 133, 134, 135, 136, 139, 140, 142, 144, 145, 150, 154, 157, 159, 161, 162
testing, 4, 6, 37, 40, 92, 98
therapy, 96, 98, 102
thermal analysis, 79, 85
thermodynamic method, 82, 86
thermodynamic properties, 72, 73, 90
thermodynamical parameters, 114
thermodynamics, 131, 132
thermograms, 76, 77, 79
thrombus, 105
tissue, ix, 23, 25, 29, 31, 32, 35, 79, 86, 88, 89, 96, 98, 100, 103, 109, 113, 114, 115, 116, 117, 118, 124, 125, 131, 133, 142
topology, 110
toxicity, 21, 22

transformation, 34, 63
transfusion, 92, 98, 100
transition temperature, 68, 75, 77
transmission electron microscopy, 29
transplantation, ix, 92, 93, 98, 100, 102, 109
transport, ix, 76, 78, 79, 80, 85, 87, 88, 89, 91, 93, 94, 102, 107, 111, 112, 113, 173, 174
treatment, 4, 21, 24, 26, 34, 40, 102, 105
tumor, 79, 88

U

umbilical cord, 95, 103, 105
universal gas constant, 76
USA, 50, 92, 98
uterine fibroids, 79

V

vacuum, 7, 9, 13, 14, 94, 96
validation, 46, 171
vapor, 15, 29, 35
variables, 141, 143
vector, 136, 150, 163, 169
vegetables, 41
velocity, 150, 161, 162, 163, 168, 169
vessels, 142
viruses, 6
viscosity, 73, 74, 86, 89, 122, 146, 147, 163, 165, 168, 174
visualization, 27
vitamin E, 101

W

Washington, 56
water, 4, 7, 10, 11, 12, 14, 15, 17, 19, 20, 23, 24, 26, 30, 37, 39, 42, 43, 49, 51, 52, 72, 73, 74, 76, 77, 78, 79, 80, 81, 82, 84, 85, 86, 87, 88, 89, 90, 95, 96, 97, 111, 112, 113, 114, 131, 132, 133, 134, 135, 136, 137, 138, 140, 141, 143, 144, 145, 159, 161, 163, 172
water permeability, 88
Western Australia, 66
Wisconsin, 98, 107
wood, 10, 13, 15, 37, 49, 50, 51, 52, 53, 64

Y

yeast, 14, 16, 22, 61, 84, 85, 90, 94
yield, 34, 47, 60, 144